Hydraulic Engineering Circular No. 24

HIGHWAY STORMWATER PUMP STATION DESIGN

Technical Report Documentation Page

1. Report No.	2. Governmental Accession No.	3. Recipient's Catalog No.

4. Title and Subtitle	5. Report Date
Highway Stormwater Pump Station Design Hydraulic Engineering Circular 24	February 2001
	6. Performing Organization Code

7. Author(s)	8. Performing Organization Report No.
Peter N. Smith	

9. Performing Organization Name and Address	10. Work Unit No. (TRAIS)
Parsons Brinckerhoff Barton Oaks Plaza Two 901 Mopac Expressway South, Suite 595 Austin, TX 78746	
	11. Contract or Grant No. DTFH61-97-C-00033

12. Sponsoring Agency Name and Address	13. Type of Report and Period Covered
Federal Highway Administration Office of Bridge Technology Applications, HIBT-20 400 Seventh Street, SW., Room 3203 Washington, D.C. 20590	Final Report
	14. Sponsoring Agency Code

15. Supplementary Notes

FHWA COTR: Arlo Waddoups (Ayres Associates) formerly of FHWA
Technical Assistance: Philip Thompson, Johnny Morris (Ayres Associates), Dan Ghere (Illinois DOT),

16. Abstract

This circular provides a comprehensive and practical guide for the design of stormwater pump station systems associated with transportation facilities. Guidance is provided for the planning and design of pump stations which collect, convey, and discharge stormwater flowing within and along the right-of-way of transportation systems. Methods and procedures are given for determining cumulative inflow, system storage needs, pump configuration and selection, discharge system size, and sump dimensions. Pump house features are identified and construction and maintenance considerations are addressed. Additionally, considerations for retrofitting existing storm water pump stations are presented.

17. Key Words	18. Distribution Statement
Pump Station, Pumps, Storm Drain, Stormwater Management, Hydraulic, Hydraulic Engineering Circular, HEC, Highway Hydraulic Design	This document is available to the public from National Technical Information Service, Springfield, Virginia 22151

19. Security Classif. (of this report)	20. Security Classif. (of this page)	21. No. of Pages	22. Price
Unclassified	Unclassified	218	

Form DOT F 1700.7 (8-72)Reproduction of completed page authorized

This page intentionally left blank.

Table of Contents

Table of Figures

Table of Tables

List of Symbols

Symbol	Description	Units, S.I. (English)
A	area of flow at uniform depth	m^2 (ft^2)
A_S	cross-section area of conduit	m^2 (ft^2)
A	distance from centerline of pump inlet bell/volute to sump entrance	m (ft)
a	length of constricted bay section at pump	m (ft)
B	breadth (horizontal plane)	m (ft)
BP	brake power	kW (hp)
B	clearance from back wall to centerline of pump inlet bell/volute	m (ft)
C_u	unit conversion coefficient/factor	--
C	unit coefficient/factor	--
C	clearance between pump inlet bell and sump floor	m (ft)
C_b	minimum clearance between adjacent bells/volutes	m (ft)
C_w	minimum clearance between bell/volute and closest sump wall	m (ft)
D	diameter or rise	m (ft)
D	pipe diameter	m (ft)
D	outside diameter of pump bell/volute	m (ft)
D_p	diameter of inflow conduit	m (ft)
D_s	inside diameter of sump	m (ft)
d	deepest depth in element	m (ft)
f	friction factor	--
g	acceleration due to gravity	m/s^2 (ft/s^2)
H	pump head at BEP	m (ft)
H_{pa}	atmospheric pressure head on the surface of the liquid in the sump	m (ft)
H_p	pressure head change between outlet and intake	m (ft)
H_s	static head	m (ft)
H_s	static suction head of liquid	m (ft)
H_{vp}	the vapor pressure head of the liquid at the operating temperature	m (ft)
H_{min}	minimum water depth in sump	m (ft)
h	minimum height of constricted bay section	m(ft)
h_f	friction head loss	m (ft)
h_l	friction loss through appurtenance	m (ft)
h_V	velocity head	m (ft)
Σh_f	total friction losses in the suction line	m (ft)
K	loss factor based on standard data or manufacturer's specified data	--
L	length of pipe	m (ft)
L	length of element	m (ft)
$L_{s(req)}$	total length of storage unit required	m (ft)
L_p	unobstructed straight length of inflow conduit	m (ft)
N	pump speed	rpm
N_{ss}	suction specific speed	rpm
N	number of pumps in the station	--
$NPSHR$	net positive suction head required	m (ft)
n	Manning's roughness coefficient	--

List of Symbols (Continued)

Symbol	Description	Units, S.I. (English)
Q	discharge in pipe	m^3/s (cfs)
Q	pump capacity at Best Efficiency Point	m^3/hr (gpm)
Q_i	a worst-case constant inflow rate for the particular pump cycle	m^3/s (cfs)
Q_p	individual pump rate	m^3/s (cfs)
Q_u	discharge	m^3/s (cfs)
R	hydraulic radius	m (ft)
S	slope of conduit	m/m (ft/ft)
S	minimum pump inlet bell submergence	m (ft)
s	slope of element	m/m (ft/ft)
s_s	side slope (ratio of vertical to horizontal)	m/m (ft/ft)
TDH	total dynamic head	m (ft)
t	minimum cycle time	minutes
V	volume of element	m^3 (ft^3)
V_{cs}	Volume in collection system below allowable highwater	m^3 (ft^3)
V_{min}	minimum required cycle volume	m^3 (ft^3)
V$_n$	cycle volume for nth pump	m^3 (ft^3)
V_{req}	Minimum storage volume required	m^3 (ft^3)
V_s	average velocity in pump bay	m/s (fps)
V_w	Volume in wet well below allowable highwater	m^3 (ft^3)
v	flow velocity	m/s (fps)
W	width of element	m (ft)
W	pump inlet bay width	m (ft)
w	minimum width of constricted bay section	m (ft)
WP	water power	kW (hp)
WWP	wire-to-water power	kW (hp)
X	pump inlet bay length	m (ft)
Y	distance from centerline of pump inlet bell to screen	m (ft)
Z$_1$	distance from centerline of pump inlet bell to diverging walls	m (ft)
Z$_2$	distance from centerline of pump inlet bell to sloping floor	m (ft)
α	angle of floor slope	degrees
β	angle of wall convergence	degrees
γ	specific weight of liquid	N/m^3 (lbf/ft^3)
η	pump efficiency	--
η_e	motor efficiency	--
ϕ	angle of convergence from constricted area to pump bay walls	degrees

1. INTRODUCTION

1.1 NEED FOR STORMWATER PUMP STATIONS

Stormwater pumping stations are necessary for the removal of stormwater from sections of highway where gravity drainage is impossible or impractical. However, stormwater pumping stations are expensive to operate and maintain and have a number of potential problems that must be addressed. Therefore, the use of stormwater pumping stations is recommended only where no other practicable alternative is available. Alternatives to pumping stations include siphons, recharge basins, deep and long storm drain systems and tunnels.

1.2 INTENT OF MANUAL

This manual is intended primarily for highway drainage designers and others interested in the hydraulic design of highway stormwater pump stations. Though some discussion relates to other engineering disciplines and responsibilities, the information is basic and intended only to enhance the hydraulic designer's ability to accommodate other needs and communicate with designers from other disciplines.

1.3 ORGANIZATION OF MANUAL

This manual is divided into fourteen chapters including this introduction. The general organization can be classified as follows:

1. Identification and basic concepts (Chapters 2 and 3)
2. Design process (Chapter 4)
3. Design criteria, considerations, and procedures (Chapters 4 through 9)
4. Additional Information (Chapters 10 through 14, and appendices)

1.4 UNIT CONVENTION

The general convention employed in this manual is to present values and dimensions in System Internationale (SI) units followed by English units in parentheses. Where practicable, the manual provides equations with unit conversion factors. In this manner, only one equation appears for a particular operation and the user must select the desired units and unit conversion factor.

1.4.1 SI versus Metric

System Internationale (SI) units are very specific. Not all metric units are SI. For example, linear measurements of millimeters and meters are metric and SI, whereas centimeters are metric but not SI. This manual uses SI units except where noted to conform to industry standards.

1.4.2 Caution on Unit Usage

Most manufacturers in the US develop pumps and pumping equipment in English units. Few present design data in SI units. The designer should take care when using and quoting units because some variables that are seemingly dimensionless may have units. Some commonly used coefficients have dimensions, which, in the strictest sense, should take on different values when using SI units. These will be noted where appropriate throughout the text.

2. PUMP STATION COMPONENTS

2.1 INTRODUCTION

This chapter identifies the major components of a highway stormwater pump station. Figure 2-1 shows a pump station and its major components.

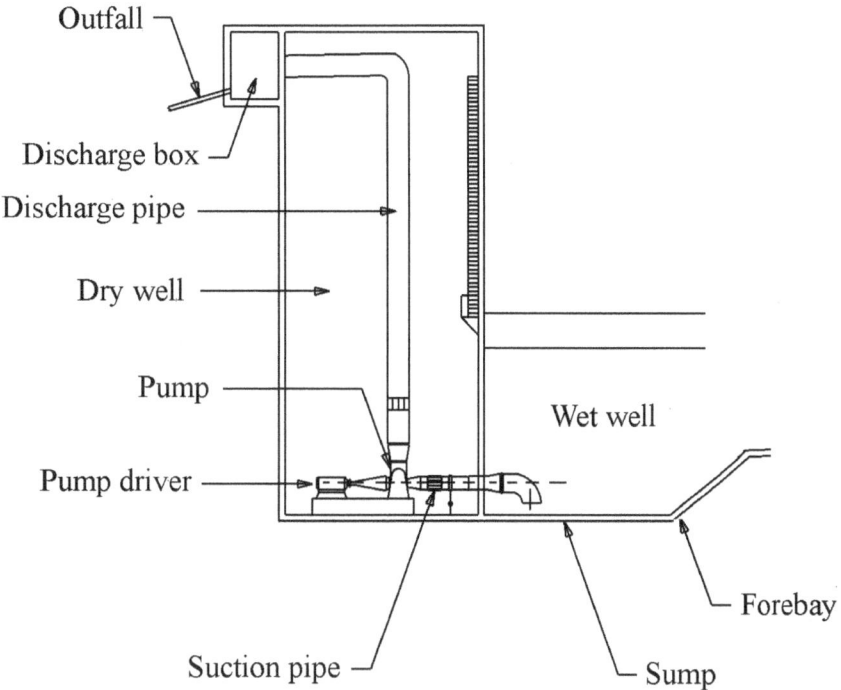

Figure 2-1. Highway stormwater pump station

2.2 COLLECTION SYSTEM

Water is conveyed from the drainage area and highway to the pump station in a system of ditches, gutters, inlets, and conduits that comprise a collection system.

2.3 STORAGE UNIT

Many systems are designed to include an underground storage area using box or circular conduits.

2.4 WELLS

All stations comprise a chamber (wet well) from which the stormwater is pumped. Stations in which the pumps are placed in the wet well are known as wet-pit stations. Some stations use a separate well (dry well) to house the pump and driver and are referred to as dry-pit stations.

2.5 PUMPS

Pumps are devices that increase the static pressure of fluids. In other words, pumps add energy to a body of fluid in order to move it from one point to another. The increase in static pressure can be achieved in different ways. Chapter 3 describes various pumps by which this result can be achieved and the characteristics of such pumps.

2.6 PUMP CONTROLS

The pump controls activate the pumps successively in response to a rising water level in the sump. The controls regulate pump activity until the inflow into the wet well has been evacuated.

2.7 PUMP DRIVERS

Electric motors and diesel, liquid propane gas, or natural gas engines are used as pump drivers at stormwater pump stations. Alternating-current electric motors are the most commonly used type of driver. Engines are used infrequently because of the noise and vibrations they produce.

2.8 STATION HOUSING

Figure 2-2 shows a photograph of a typical pump house for a highway stormwater pump station. General features of a pump house include:

- ventilation,
- doors and hatches,
- stairs and ladders,
- cranes and hoists, and
- safety and security measures.

Figure 2-2. Highway stormwater pump house

2.9   SUMP PUMP

A sump pump is sometimes installed in the sump of wet wells of wet-pit and dry-pit stations. The sump pump is used to pump out the water remaining in the well after the water level has dropped and all the pumps are no longer pumping.

The sump pump may also remove the accumulation of solids, such as silt, sand and debris that builds-up gradually at the bottom of the wet well.

2.10   POWER SUPPLY

Electric power is usually the most economical and reliable power source if it is available. Diesel, liquid propane, and natural gas are also usually available to drive engines if no electrical service is available or if there is an interruption in service.

Pumping stations also require a backup power source. The need or requirement for standby power will vary with each installation and the decision should be based on economics and safety. The three principal types of backup power usually considered are multiple electric feeders, engine-driven pumping units and electric generators.

2.11   SECURITY AND ACCESS FEATURES

2.11.1   Security

Pump stations usually include features to secure them from entry by unauthorized personnel and to minimize the risk of vandalism. This may be achieved by including as few windows as possible in the design, providing fencing and exterior lighting, concealing expensive equipment and installing unauthorized entry alarms.

2.11.2 Access

Typical access features include:

- off-highway driveways or service roads,
- parking, and
- loading zones.

2.12 OTHER FEATURES

2.12.1 Hoists

Hoists are usually placed over basket screens, submersible pumps, other pumps or motors, and other locations where it is necessary to lift heavy pieces of machinery or equipment.

2.12.2 Trash Racks

Trash racks may be provided at the entrance to the wet pit in order to prevent large debris from entering the wet well.

3. PUMPS AND PUMP STATION TYPES

3.1   INTRODUCTION

This chapter identifies the types of pumps and pump stations that are used for highway stormwater drainage.

3.2   CLASSIFICATION OF PUMPS

Pumps are often classified by the way in which they impart energy. The two primary classes are:

- dynamic or kinetic pumps, and
- positive displacement pumps.

3.2.1   Dynamic Pumps

Dynamic pumps are also called centrifugal pumps. The main feature of this kind of pump is that the liquid is delivered in a continuous and uninterrupted flow by means of an impeller - a rotating device with vanes. The impeller transfers energy to the liquid being pumped, partly by an increase in pressure, by producing a high liquid velocity greater than the velocities occurring at discharge, and partly by an increase in kinetic energy that is later converted into pressure energy.[1]

Dynamic Pumps may be divided into several varieties of centrifugal and other special effect pumps (vertical shaft pumps). A common feature of centrifugal pumps is that they are equipped with a volute or casing.

3.2.2   Positive Displacement Pumps

Positive displacement pumps are those in which the moving element of the pump forces a fixed volume of fluid from the inlet pressure section of the pump into the discharge zone of the pump resulting in an increase in the pressure of the liquid. Such pump types are seldom used for highway stormwater pump stations and are not discussed further.

3.3   IMPELLER TYPES

The three typical impeller types for pumps are axial, radial and mixed. A description of each of these flow-types follows. Refer to Figure 3-1 for identification of each type.

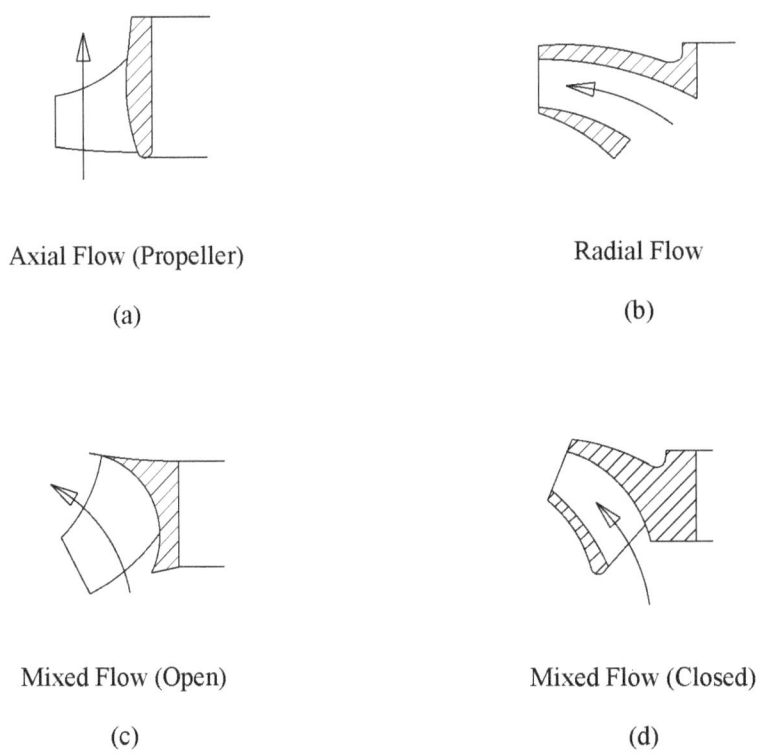

Axial Flow (Propeller)

(a)

Radial Flow

(b)

Mixed Flow (Open)

(c)

Mixed Flow (Closed)

(d)

Figure 3-1. Impeller types

3.3.1   Axial-Flow Pumps

Pumps of the axial-flow type have impellers that lift the water up a vertical riser pipe and direct the flow axially through the impeller parallel to the pump axis and drive shaft (Figure 3-1. a). Horizontally mounted, axial-flow pumps are high-capacity pumps that are typically used for low head, high discharge applications such as flood control.

3.3.2   Radial-Flow Pumps

Radial (centrifugal) flow pumps have impellers that force the liquid to discharge radially, from the hub to the periphery, by using centrifugal force to move water up a riser pipe. Impellers of this type are called radial flow or radial vane impellers (Figure 3-1. b). The pumps handle any range of head and discharge, but are most suited to high head applications.

3.3.3   Mixed-Flow Pumps

The mixed-flow type has impellers with vanes which are shaped such that the pumping head is developed partly by centrifugal force and partly by a lifting action of the vanes. They are either open type or closed type (Figure 3-1. c and Figure 3-1. d respectively). The pumps are used primarily for immediate head and discharge applications.

3.3.4 Propeller versus Impeller

Propeller pumps are used for low-head pumping such as is typical for highway stormwater pumping stations. The propellers are generally of the vertical single-stage, axial- and mixed-flow type. Two-stage, axial-flow pumps may also be used. Propellers are usually referred to as mixed-flow impellers with a very small radial-flow component. A true propeller, in which the flow strictly parallels the axis of rotation, is called an axial-flow impeller.

Impellers are called radial vane or radial. As the impeller turns, the centrifugal action creates a vacuum at the impeller "eye" and causes the fluid to be discharged radially to the periphery with increasing velocity. The high velocity at the blade tips is converted to pressure in the casing. A higher impeller speed results in a higher blade velocity, which means higher pressures. The impellers of centrifugal pumps may be of the radial- or mixed-flow type.

3.3.5 Comparison of Impeller Types

The following section lists some advantages and disadvantages associated with different impeller types:

Advantages
- Radial flow pumps with a single vane, non-clog impeller handle debris the best because of the large impeller opening.
- Mixed-flow pumps deliver the liquid at a higher head than the propeller pump.

Disadvantages
- Debris handling capability of radial-flow pumps decreases with an increase in the number of vanes.
- The propellers of axial-flow pumps are susceptible to damage from large, hard objects.

3.4 PUMPS FOR STORMWATER USE

3.4.1 Submersible Pumps

Submersible pumps are close-coupled pumps driven by a submersible motor and are generally a vertical type of pump. One vertical type of submersible pump (Figure 3-2) has a completely recessed impeller, out of the stream of pumped liquid, while another has a specially shaped non-clog impeller. The in-flow, non-clog impeller has substantially higher efficiencies than the recessed impeller.

Submersible pumps require that the casing of the pump, which provides the suction and causes the material to flow into the impeller blades, be submerged in the liquid in order to pump. If the top cover is above the level of the liquid then air will enter the casing and prevent liquid from being discharged. When the casing is submerged, the pump will self-prime and start pumping. Submersible pumps are designed to be immersed in water and so are particularly suitable for wet well installation and for use as sump pumps.

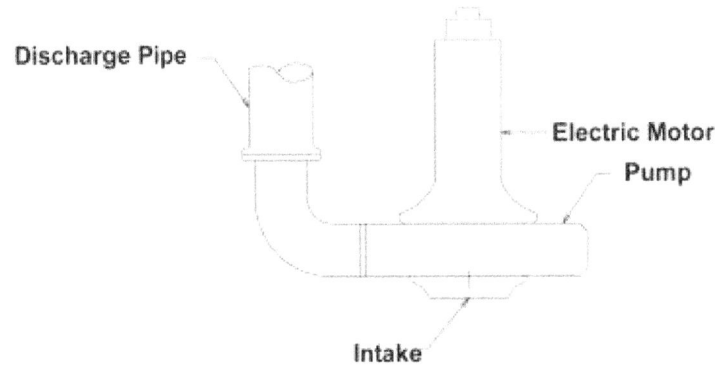

Figure 3-2. Submersible pump

3.4.1.1 Advantages of Submersible Pumps

Some of the advantages of using submersible pumps include:

- available size of submersible pumps, making it possible to design large stormwater stations with sole reliance on submersible pumps,
- wide range of sizes and characteristics for a vertical propeller pump,
- protection against dry well flooding,
- easy removal for repairs,
- elimination of need for long drive shafts,
- natural cooling by stormwater, and
- shorter allowable cycle times.

3.4.1.2 Disadvantages of Submersible Pumps

Some of the disadvantages of using submersible pumps are:

- limited availability of motor sizes,
- more expensive pump and motor,
- need to operate in submerged conditions for cooling, and
- potential need for larger wet well size and pump spacing.

3.4.2 Vertical Shaft Pumps

Vertical shaft pumps, heretofore referred to as vertical pumps, are a commonly used type of pump for stormwater pumping. The pumps are equipped with an axial diffuser (or discharge bowl) rather than a volute. The pump assembly of vertical pumps is suspended below the baseplate by the necessary length of discharge column and elbow, with the pump bowl being adequately submerged into the liquid being pumped (Figure 3-3).

The bowl assembly is the heart of the vertical pump and includes a suction bell, discharge bowl, impeller, pump shaft, bearings and the parts necessary to secure the impeller to the shaft. The suction bowl is designed to permit proper distribution of the liquid to the impeller. Vertical pumps for highway stormwater pumping stations will usually be the single-stage, axial-flow type or the mixed-flow type.

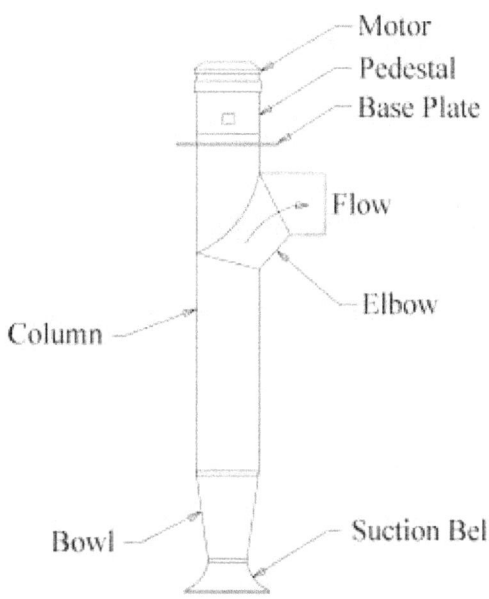

Figure 3-3. Vertical shaft-driven pump

Typically, single-stage propeller pumps are used for low heads and mixed-flow pumps for higher heads. A two-stage propeller pump approximately doubles the head capacity of a single stage.

3.4.2.1 Advantages of Vertical Pumps

The advantages of selecting a vertical shaft pump for stormwater pumping include:

- low maintenance requirements,
- easy access to the driver,
- use of drive shafts up to 12-m (40-ft),
- freshwater flushing of bearings as a viable lubrication alternative to grease where there is a municipal water supply, and
- a small floor area which is sufficient to accommodate the axial pump bowl.

3.4.2.2 Disadvantages of Vertical Pumps

The disadvantages of selecting a vertical shaft pump for stormwater pumping include:

- questionable cost-effectiveness of water lubrication (because of the small number of annual operating hours of a stormwater pumping station and the relative ease of providing effective grease lubrication),
- unsuitability of shallow sumps (due to submergence requirements),
- requirement for two additional elbows for the vertical type, angle-flow pump -- one in the suction and one in the discharge,
- limited access to the pump, which requires pulling the entire assembly out of the pump station,
- noise level, which is generally greater than for submersible pumps, and
- shaft stretch, which wears impellers and bowls unless the thrust bearing is kept carefully adjusted.

3.4.3 Horizontal Pumps

A horizontal centrifugal (non-clog) pump is shown in Figure 3-4. Either an electric motor or engine drives a horizontal shaft on which the pump impeller is mounted. Usually, the impeller is housed in a spiral-shaped casing called a volute. The unit is either frame-mounted or close-coupled with the motor on the floor of a dry chamber (dry well). Stormwater is collected in a separate chamber (wet well) and is introduced into the pump through a suction pipe.

For highway pump stations, use of a horizontal pump usually implies suction lift meaning that the intake pipe is lower than the centerline of the axis of the impeller. Water is taken in along the impeller axis and ejected perpendicular and vertical to the impeller axis.

Figure 3-4. Horizontal pump

3.4.3.1   Advantages of Horizontal Pumps

The advantages associated with using a horizontal centrifugal pump at a stormwater pumping station include:

- high reliability,
- mechanical and electrical simplicity of the horizontal centrifugal non-clog pump compared with the vertical shaft pump,
- easy access for maintenance, since the pumps are placed in a dry well,
- longer service life than vertical pumps,
- less costly motors than for vertical pumps, and
- lower headroom requirements than vertical pumps.

3.4.3.2   Disadvantages of Horizontal Pumps

Some of the disadvantages associated with using a horizontal centrifugal pump at a stormwater pumping station are that:

- the impellers typically have an efficiency of 70% or lower,
- a dry well is required,
- a large pump room floor area is required,
- the motors are subject to flooding if the dry well becomes inundated, and
- ventilation is needed to cool motors.

3.5   WET-PIT STATIONS

A wet-pit station typically comprises a wet well or several wet wells and a pump house. Generally, one of two configurations is used: rectangular or circular caisson. Two types of pumps, vertical and submersible, may be installed in the wet well of the wet-pit station.

Vertical pumps are usually installed in either rectangular wells (Section 3.5.1) or in circular wells (Section 3.5.2). Submersible pumps are commonly installed in rectangular pits (Section 3.5.3). Simple rectangular or circular wells are used for the smaller submersible pumps (Section 3.5.4).

3.5.1   Vertical Pumps in Rectangular Wells

Figure 3-5 shows a simplified diagram of an installation of vertical pumps in a rectangular well of a wet-pit station. Figure 3-6 shows a typical sketch of an installation used by Arizona DOT.

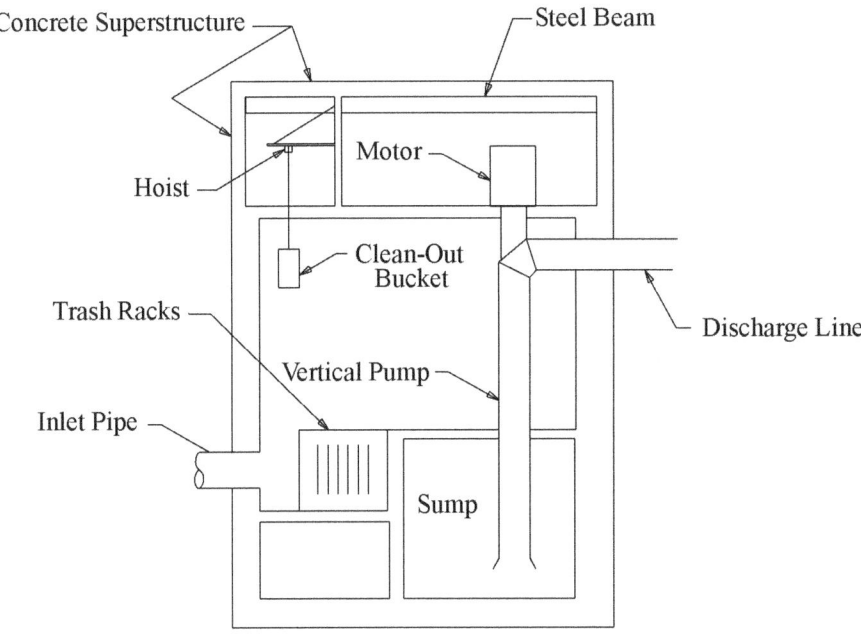

Figure 3-5. Vertical pump in rectangular wet well

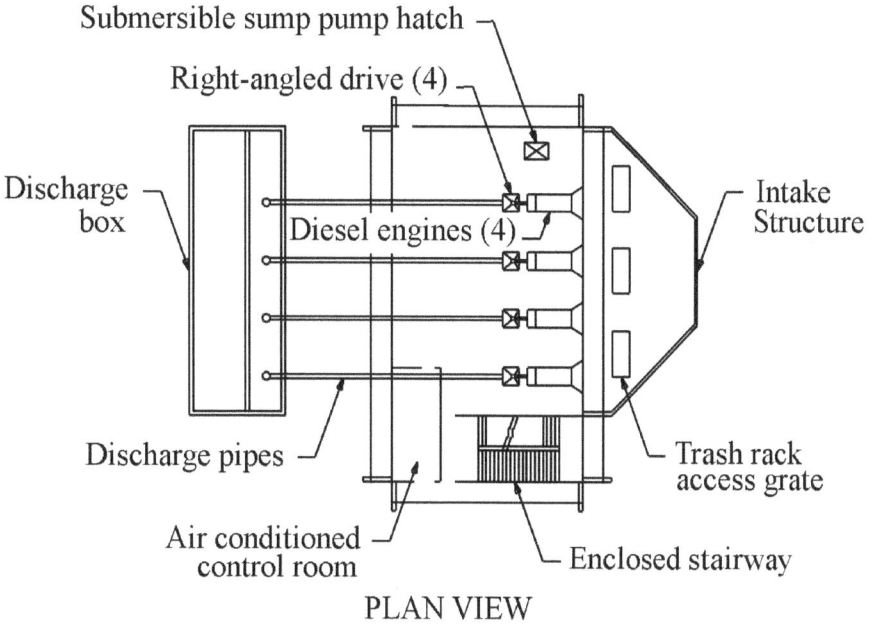

PLAN VIEW

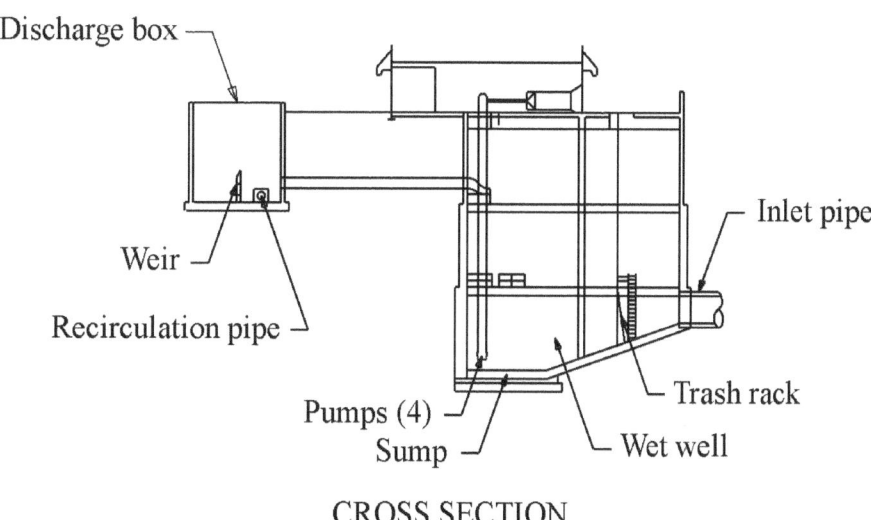

CROSS SECTION

Figure 3-6. Vertical pumps in rectangular well used by Arizona DOT.

3.5.1.1 Alignment

Vertical pumps within a rectangular well are placed in a straight line perpendicular to the direction of the flow, with a certain minimum spacing between the pumps and back wall, depending on pump capacity and bell diameters. Figure 3-7 shows the recommended alignment. For criteria on well dimensions, refer to Section 9.1 - Sump Size and Clearance Criteria.

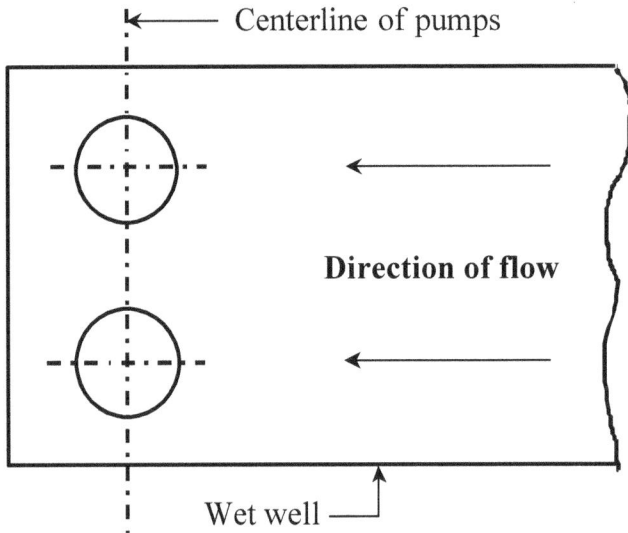

Figure 3-7. Alignment of pumps in rectangular well

3.5.1.2 Drive Shafts

The pumps are available with two different types of drive-shaft construction and lubrication. These are: 1) pumps with oil-lubricated line shaft bearings in an enclosing tube, which are the preferred choice, and 2) pumps with open shafts and grease-lubricated bearings.

3.5.1.3 Trash Racks

When a trash rack is placed in the wet well, it is necessary to ensure that there is an adequate distance between the trash rack and the back wall of the wet well to ensure smooth flow to the pumps. Refer to Section 9.1 - Sump Size and Clearance Criteria for recommended criteria.

3.5.1.4 Obstructions and Streamlining

A streamlined flow should be maintained throughout the length of the pit. Generally, stairways and features other than trash racks and flow directional vanes should not be placed in the flow path to the pumps.

Some configurations may require structural supports that extend down into the sump and could interfere with the flow patterns. Such interference could induce eddying and subsequent vortexing. To minimize the potential for eddying, any necessary obstructions should be streamlined and orientated to direct flow to the pumps.

3.5.1.5 Submergence

The depth of the sump is a function of the required submergence. Suction umbrellas (Figure 3-8) may be used to reduce the submergence requirements. The umbrella provides a smooth transition for flow entering the pump bowl. This helps reduce the potential for vortexing.

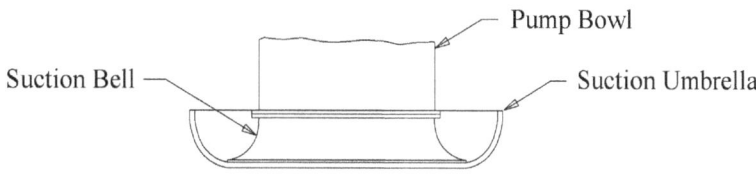

Figure 3-8. Typical suction umbrella

3.5.1.6 Access

Provision must be made for access ladders, access holes, power operated crane hoists and disposal of debris. Access should also be provided to the area around the pump intakes for clearing and maintenance.

3.5.1.7 Provisions for Sediment-Laden Water

For rectangular wells, provisions to reduce potential problems with sediment include:

- designing the wet well to provide velocities of 1 m/s (3 ft/s) or higher – though such an approach may require modeling and other provisions to reduce the potential for vortexing.
- providing for violent mixing to suspend sediments (and removed by the main pumps). Measures include using submerged mixers and bypassing part of the pump discharge back into the well.
- dewatering and pressure-cleaning the well.
- designing a confined wet well. Refer to 9.1.4.6 Accommodation of Solids-bearing Water in Rectangular Wells.

3.5.1.8 Sump Pump

A sump pump is sometimes placed in wet wells and dry wells. Submergence and clearance requirements will usually result in at least 1 m (3 ft) depth of water remaining in the sump after the main pumps cease. Some state highway agencies prefer to evacuate this water and any sedimentation to help prevent clogging, excessive wear, and corrosion of the main pumps.

3.5.1.9 Construction of Rectangular Well

The most common construction technique for a rectangular well is open cut with sheet piling.

3.5.1.10 Pump House

The housing is usually above the wet well to enclose the pumps and drivers and house the controls. Figure 3-9 shows a sample of a small pump house enclosing two vertical pumps in a rectangular wet well.

Figure 3-9. Small pump house for vertical pumps in Houston, Texas

3.5.2   Vertical Pumps in Circular Wells

3.5.2.1   Alignment

When placed in a circular caisson, the pumps should be placed with their centers along the center of the sump or offset from the center of the sump on the opposite half of the semicircle to the intake line. Generally, a maximum of three pumps may be used in any one sump. For criteria on well dimensions and configurations, refer to Section 9.1.5 - Circular Sumps.

3.5.2.2   Driver Accommodations

Both electrically and engine-driven pumps may be used in circular wells. When engines are used as drivers, the engines are housed in a separate aboveground rectangular structure due to the limited space within a circular caisson.

3.5.2.3   Submergence

The effectiveness of the layout and pump operation depends largely on supplying ample submergence for the pumps. Submergence requirements should be met by deepening the caisson.

3.5.2.4   Trash Racks

If trash racks are provided, they may be placed at the inlet side of the sump or in a forebay or separate well. The trash racks should be easily assessable due to the relatively small trash and debris storage capability provided in the caisson.

3.5.2.5 Housing

As with vertical pumps in rectangular wells, the pump house encloses the motors and other station equipment above the wet well.

3.5.2.6 Access

Provision must be made for access ladders, access holes, power operated crane hoists and disposal of debris. There should also be access to the area around the pump intakes for clearing and maintenance.

3.5.2.7 Provisions for Sediment-Laden Water

A quantity of the sand and silt that enters the station through the pit inlets passes immediately through the pumps. For circular wet wells, provisions to reduce potential problems with sediment include:

- providing a separate well with sump pump with a lower floor than the main pumping well(s) to collect sediment,
- using the storage system as a stilling area and making provision for removing the sediment, and
- sloping the walls and minimizing horizontal floor space. Refer to 9.1.5.5 Accommodation of Solids-bearing Water in Circular Wells.

In addition, for vertical pumps it will be preferable to provide a pump lineshaft enclosing tube packed with a light hydraulic grease and individual grease lines for each pump bowl bearing, with grease seals to limit entrance of silt.

3.5.3 Submersible Pumps in Rectangular Wells

Rectangular pump pits provide the most favorable conditions for flow into and through the pit for submersible pumps. Large pumps can be considered to have an individual capacity of greater than 680 m^3/hour (3000 gpm). The large vertical submersible pumps used in the rectangular pit configuration may have as many as ten starts per hour. The relatively high allowable number of cycles results in a lower required storage volume than required for vertical pumps. Figure 3-10 shows a rectangular pit in which submersible pumps are installed. The pumps are hung inside plastic piping and can be lifted from access holes above the wet well. Divider walls help distribute the inflow and reduce the potential for vortex formation. Figure 3-11 shows the submersible pumps prior to installation.

Figure 3-10. Rectangular pit with submersible pumps

Figure 3-11. Large submersible pumps

3.5.3.1 Housing

No elements of the wet well need appear above ground, so it is common to provide a separate pump house for the controls, generator, and other equipment. Alignment, flow distribution and clearance considerations are similar to those described in Section 3.5.1 for vertical pumps in rectangular wells. It is common and desirable to use an inlet chamber.

3.5.3.2 Inlet Chamber

An inlet chamber prevents the inflow from directly splashing into the pump chamber and conveying air along with it. An overhang or hanging baffle at the inlet of smaller pipes may be selected to prevent the incoming water from directly splashing into the pump chamber. For larger systems, the intake structures should be designed with adequate width and depth to limit the maximum pump approach velocities to 0.5m/s (1.5 ft/sec). The horizontal axis of the inlet structure should be perpendicular to the horizontal axis of the pumps and be positioned to ensure even distribution of flow to the pumps without incurring eddies.

3.5.4 Submersible Pumps in Circular Wells

Normally, there should be a minimum of two pumps installed in the well, even though a single pump may provide the desired capacity. Figure 3-12 shows a sample layout in which submersible pumps are placed in several circular wet wells. The first wet well (Figure 3-13) acts as a sump for a sump pump and the other two wells house the main pumps.

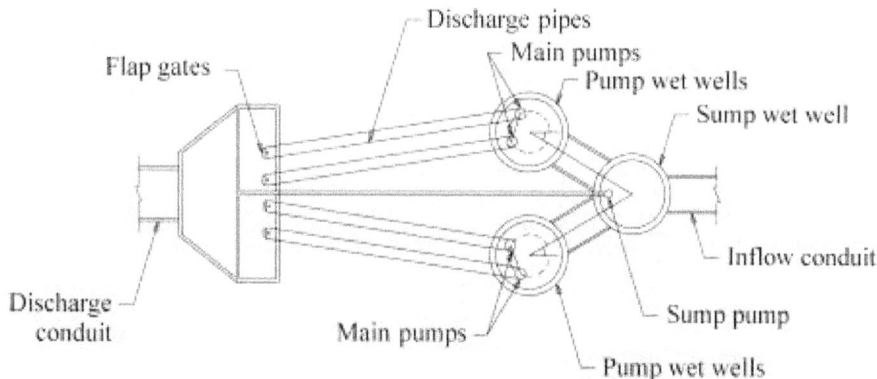

Figure 3-12. Sample layout for submersible pumps in circular caissons

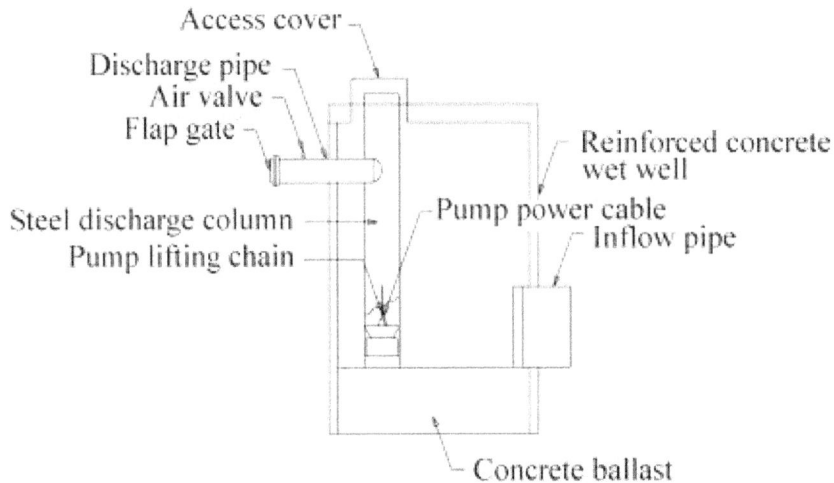

Figure 3-13. Circular wet well with submersible pumps – elevation

3.5.4.1   Construction

A primary advantage of circular caissons is the efficiency with which they can be constructed. Typically precast units are aligned and sunk into the ground. Additional sections may be added as the first section is lowered. The base is filled with in-situ concrete to provide ballast as well as the sump floor. Figure 3-14 shows caissons during construction.

Figure 3-14. Circular caisson wet wells under construction.

3.5.4.2 Storage Requirements

Small submersible pumps usually have a capacity of 680 m³/hr (3000 gpm) or less. Generally, for small submersible pumps, cycle times can be as low as 2 minutes, resulting in significantly smaller system storage requirements than other types of pump. Similarly, larger submersible pumps have shorter minimum cycle times than many other pumps because of the cooling resulting from the stormwater. This makes their use preferable for situations in which there is little or no provision for storage other than that provided by the wet well. However, for the larger systems, storage outside the wet wells will reduce the size of the pumps and is typically cost effective.

3.5.4.3 Pump Elevations

Some agencies may set the pumps at different elevations within the caisson not only to provide for low-flow conditions but also to guard against the build-up of sand or silt in the bottom of the pit, thereby reducing the risk of damage to all the pumps. However, the pump set at the lowest elevation may experience excessive wear because it must always be the first and last to operate. Therefore, a preferred approach is to set all the main pumps at the same elevation and add a pump of smaller capacity than the main pumps.

3.5.4.4 Pump Configuration

Submersible pumps usually have bottom suction and side discharge. The motor is commonly mounted vertically above the volute. The impeller is then directly mounted on the motor shaft extension. The motor also may be mounted horizontally, in which case water is drawn axially past the motor and through the propeller into the discharge line.

3.5.4.5 Housing

Housing may not need to be above ground if controls can be placed in a confined space with minimal spacing.

The arrangement of the pumps in circular caissons makes the removal of any pump for maintenance or inspection easy. Removal becomes necessary especially when the build-up of sand and silt reaches sufficient quantity. The accessibility of the pumps makes the clean out of the station a relatively simple operation. An agitating system may be provided for a simple pit or caisson to ensure that the sump is cleaned. The agitating system involves piping water into the caisson and installing a spray ring system to agitate any sand and silt prior to starting the pumps.

Sediment accumulation can be reduced by using sloping walls and minimizing the amount of horizontal floor area.

3.5.4.6 Sump Pump

Some agencies prefer to place a sump pump either at the lowest point of the main pump wet well or in a separate sump chamber. The layout shown in Figure 3-12 has a separate wet well that is used as a sump. Figure 3-15 shows a sample detail of a sump pump in a circular caisson.

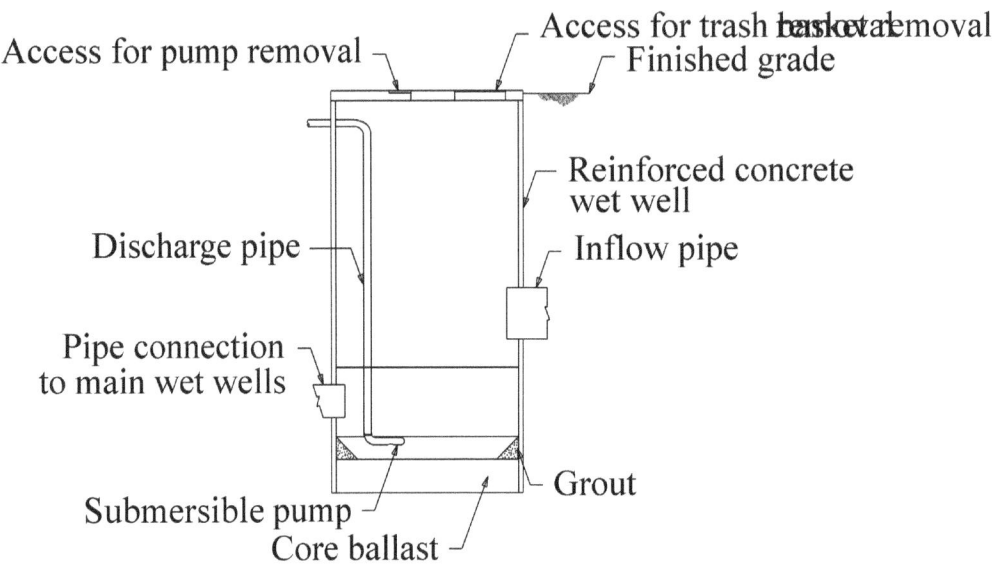

Figure 3-15. Circular caisson sump with submersible pump

3.6 DRY-PIT STATIONS

The dry-pit station comprises two main chambers: a dry well and a wet well. The stormwater enters the station and is stored in the wet well. The wet well is usually separated from the dry well by a wall and is connected to the dry well by a suction line for each pump. The stormwater pumps and controls are located on the floor of the dry well. Power to the pumps is provided by either close-coupled motors in the dry well or long drive-shafts with the motors located overhead.

3.6.1 Dry Well

The area of the dry well is usually rectangular in shape and the size depends on the number of pumps installed in the well. The standard design usually calls for a minimum of two pumps in series. Pumps are usually of the horizontal centrifugal non-clog type. This type of pump is suited for stormwater applications because it is resistant to wear from abrasion. The pumps are oriented to provide both a clockwise and counter-clockwise rotation to permit a symmetrical pumping arrangement. The small size of the centrifugal pumps allows close spacing of the pumps. Another advantage is the compact layout of the dry well, of which only a small portion is occupied by the access shafts and switchgear.

Dry wells for dry-pit stations with vertical centrifugal angle flow pumps, or vertical shaft centrifugal volute-type pumps as they are also called, are usually more costly and require greater maintenance than stations with horizontal shaft pumps. The reason for the added expense and increased maintenance is the long drive shaft required between the pump and the motor. The shaft is commonly a source of trouble due to vibration.

3.6.2   Wet Well for Dry-Pit Station

Wet wells for dry-pit stations are usually sized only large enough to meet minimum sump clearance criteria (See Section 9.1 - Sump Size and Clearance Criteria). The storage requirement for pump cycling criteria is met by designing a storage unit or larger collection system. The intake to the wet well should be designed to ensure even distribution of flow to all the pumps. The bottom of the wet well is often dished in order to direct the water to the suction line.

3.6.3   Storage Unit

The storage unit for this arrangement is usually a multiple-barrel culvert (Figure 3-16) and usually connects to the wet well without obstruction. The unit is set to a recommended slope of 2 percent to minimize the potential for settlement of solids. The pumps effect an almost complete emptying of the storage unit and sump with this configuration.

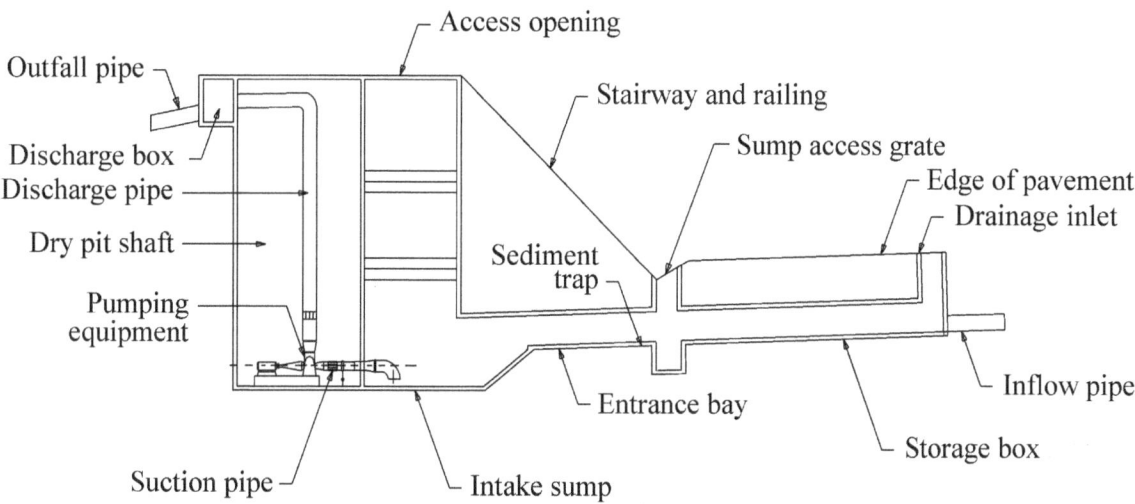

Figure 3-16. Typical storage unit and dry-pit station

3.7   COMPARISON OF PIT TYPES

The following sections cover the advantages and disadvantages of wet-pit and dry-pit stations.

3.7.1   Dry-Pit Station

3.7.1.1   Advantages of Dry-Pit Station

The advantages of selecting a dry-pit station include:

- availability of a dry area for personnel to remove pumps or perform routine and emergency pump and pipe maintenance,
- ease of access for repair and maintenance, and
- protection of equipment from fire and explosion.

3.7.1.2 Disadvantages of Dry-Pit Station

Some of the disadvantages of selecting a dry-pit station are that:

- construction is more expensive than wet-pit stations, and
- they usually require a larger plan area than wet-pit stations.

3.7.2 Wet-Pit Station

3.7.2.1 Advantages of Wet-Pit Station

Advantages of selecting a wet-pit station are that:

- some configurations require less surface area than dry-pit stations,
- they are generally less expensive to construct than dry-pit stations, and
- they allow for simpler details.

3.7.2.2 Disadvantages of Wet-Pit Station

Some disadvantages of selecting a wet-pit station include:

- more limited access to pumps for maintenance than for dry-pit stations, and
- required evacuation of water in wet well for pump maintenance.

3.8 STATION TYPE SELECTION

The choice of a wet-pit station or dry-pit station design is usually one of individual or agency preference. Dry-pit stations are preferred when:

- there is a high potential for health hazards to maintenance personnel,
- there are concerns with the potential for fire or explosion, and
- ease of maintenance is a major concern.

For highway stormwater pump stations, the use of wet-pit stations is generally preferred by highway agencies. Primary reasons for this are:

- availability of reliable, high-capacity submersible pumps which do not require a dry pit,
- lower station construction costs, and
- reduced site size requirements.

Generally, the selection of station and pump type is subjective and often dependent on local preference and experience.

3.9 DEBRIS AND SEDIMENT HANDLING

Stormwater flushes the watershed and roadway before entering the collection system carrying debris and sediment. The pumps should be protected by screens to prevent passage of the larger debris and provision should be made for accommodating sediment.

Screening may be provided by grate inlets or screened curb opening inlets in the roadway gutters and ditches, or trash racks inside the collection system, storage unit, or wet well.

One or more of several measures may be necessary to accommodate sediment loads such as:

- using an automatic flushing system and non-clog sump/sludge pump to evacuate sediment,
- frequently flushing and pumping sediment in the system manually, and
- designing sloping walls and minimizing horizontal floor area so that velocities will carry sediment to the pump intakes.

Provision should be made for handling hazardous material spills. Preferably, the pump station should be isolated from the main collection system and the effect of hazardous spills by a properly designed storage facility upstream of the station. This may be a closed box below the highway pavement or adjacent to it. The closed box must be ventilated by sufficient grating area at each end. If the station collection system design includes a closed conduit system leading directly from the highway to the pump station without any open forebay to intercept hazardous fluids or vent off volatile gases, there should be a gas-tight seal between the pump pit and the motor room in the pump station.

3.9.1.1 Trash Rack

Trash rack construction should follow the design criteria and specifications listed below:

- The rack should be fabricated from A36 structural steel. It is usually designed for convenience by designing the trash rack in modules with a number of similar panels.
- A maximum clear-space between bars of less than 40 mm (1.5 in.) is common in order to prevent passage of a large solid or similar object that might damage a pump.
- The panels should be attached to the base slab with corrosion resistant stainless steel bolts and bolted at the top to the catwalk.
- The pattern of the trash rack and a slightly sloping face should facilitate cleaning of the face by raking from the catwalk when necessary.

Trash racks in the wet well or storage unit or grate inlets in the collection system should be provided to prevent large debris from entering the sump.

This page intentionally left blank.

4. PUMP STATION DESIGN PROCESS

4.1 INTRODUCTION

This chapter provides an overview of the whole process for developing a stormwater pump station design. It presents a general order and description of what measures are taken. The design considerations and specific procedures appear in Chapters 5 through 9.

The process can begin as early as during a Major Investment Study (MIS), corridor study, or route and feasibility study. This chapter identifies the following phases during which some level of effort may be required for a pump station. These phases are:

1. Planning
2. Facility Schematic Development
3. Facility Design
4. Plans, Specifications and Estimates (PS&E)
5. Construction
6. Operation and Maintenance
7. Retrofit

4.2 OVERVIEW OF PUMP SYSTEM HYDRAULIC OPERATION

The following provides an overview of the hydraulic operation of a typical stormwater pump system during a storm event to provide some quick insight and act as a prelude to a detailed discussion of the pump design process.

1. Runoff enters the collection system and is conveyed to the storage unit and wet well of the pump station.
2. For a period of time there is no outflow as the runoff is stored within the storage unit and wet well. The water level in the wet well rises and a hydraulic gradient develops based on the rate of inflow, the water level in the wet well, and the conveyance capacity of the storage unit and collection system.
3. The first pump starts when the water level in the wet well reaches a specific elevation. The pump evacuates the flow at a rate that varies with the pump characteristics and total dynamic head.
4. The inflow rate will vary as defined by the inflow hydrograph. If the pumping rate is lower than the inflow rate, the water level in the storage unit and wet well continues to rise and the volume stored in the system increases.
5. Additional pumps start at predetermined elevations as the water level rises.
6. At some point, the inflow rate will be lower than the total pumping rate. The water level in the wet well drops as water is evacuated from storage.
7. At preset elevations, as the water level drops, individual pumps are stopped and the discharge rate drops accordingly.
8. When the water level drops to the minimum level required for submergence, the last pump is stopped. The last pump off is usually the first to have been switched on, although different switching schemes are feasible.

Several observations can be made by inspecting the above process.

- The number and/or size of pumps required drops with increasing available storage volume.
- The peak outflow decreases with increasing storage.
- The time between starts of a given pump (cycle time) is a function of storage volume, pump capacity and inflow rate.
- If the pumping sequence, start and stop elevations, and pumping rates are not correct, the water level in the wet well will exceed the maximum allowable level.
- If the pumping rates are too high or the start and stop elevations are improperly set, the water will be evacuated quickly and the pumps will operate too frequently.

4.3   PUMP SYSTEM DESIGN OBJECTIVES

Where practicable, it is preferable to avoid the need for a pump station. However, when needed, pump system design is an iterative process that has the following hydraulic objectives:

- minimize inflow by keeping drainage area as small as practicable,
- remove runoff from the highway system without exceeding critical elevations for design storm events,
- comply with discharge constraints that may exist,
- balance storage volume, pump size and number, and construction cost, and
- select minimum cycle times that exceed allowable cycle times

4.4   PUMP STATION DESIGN PROCESS FLOWCHART

The flowchart appearing in Figure 4-1 maps out the phases and stages in the stormwater pump station design process. The flowchart highlights the iterative nature of pump design. For example, after the mass routing stage, it may be necessary to adjust the storage unit, resize pumps, change the number of pumps, or change the control elevations.

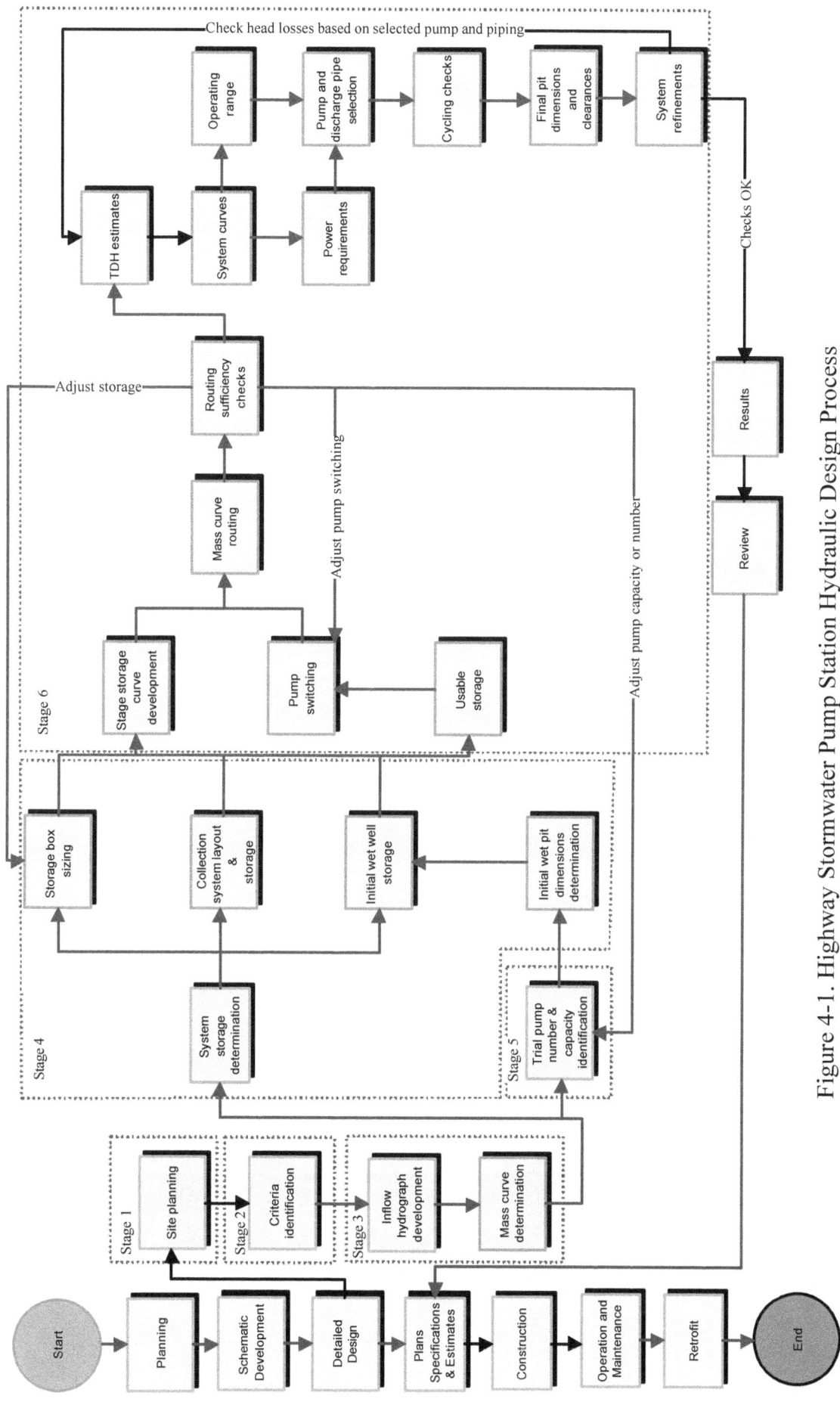

Figure 4-1. Highway Stormwater Pump Station Hydraulic Design Process

4.5   PLANNING PHASE

The planning phase includes MIS, corridor studies and other route and feasibility studies. This section provides discussion of the type and level of effort that it is reasonable to expend to determine the need for or viability of a stormwater pump station. Primarily this would be a cursory level that would help identify constraints and other bases for comparison of alternate alignments.

4.5.1   General Planning Considerations

A detailed discussion of highway planning is beyond the scope of this manual. However, the highway planners should be made aware that the potential need for a stormwater pump station should be considered when investigating alternative alignments. The high cost associated with construction, operation, and maintenance of a pump station can often render an alignment alternative undesirable.

When identifying potential highway alternative alignments, the planner should consider any feasible alternatives that would not require a stormwater pumping station. Whenever it is practicable, and other constraints notwithstanding, the preferred alternative should avoid the need for a stormwater pump station by ensuring adequate gravity drainage.

4.5.2   Level of Effort for Planning

The planning stage does not usually entail detailed engineering. However, any planning should be performed in coordination with experts from the various relevant disciplines. This should include the drainage designer. Where an alternative is likely to involve a pump station, the evaluation should include consideration of the following information if available:

- approximate drainage area required for pumping,
- approximate right of way requirements for the station,
- approximate right of way costs,
- approximate maximum static head requirements,
- average local pump station construction costs,
- average local pump operation and maintenance costs, and
- proximity to a discharge location.

An experienced pump station designer can use the drainage area and static head information to provide an approximate pump system size. Where average cost data are available, an "order of magnitude" estimate of cost can be made, allowing the pump alternative to be included in an alternatives matrix. The above discussion does not preclude the need to address environmental issues that may relate to pump stations.

4.6 SCHEMATIC DEVELOPMENT PHASE

The primary objectives of schematic development are to:

- establish a highway's vertical and horizontal alignment,
- establish typical sections, and
- identify general construction staging.

The schematic becomes the basis for the detailed design. Schematic development is a critical phase in the highway development process, after which the designer's ability to make significant changes becomes severely limited. Therefore, consideration of the potential need for a pump station should begin no later than during development of a geometric schematic.

This section discusses the considerations and level of effort appropriate for selecting a pump station. This includes:

- data collection,
- evaluation of general design criteria and constraints,
- evaluation of alignment alternatives and value engineering, and
- roadway profile guidelines.

4.6.1 Data Collection

The following data should be collected and evaluated to determine if a pump station is required and what the likely implications of vertical alignment alternatives may be:

- topographic maps/ survey data,
- outfall locations and elevations,
- approximate proposed roadway profile and horizontal alignment, and
- approximate drainage areas.

4.6.2 Evaluation of General Facility Design Criteria and Constraints

The designer should identify design criteria and typical constraints that may affect the need for a pump station. Usually, the general location of the highway is established during the planning phase. The designer will then be limited to the alternatives that are available. Typically, the following factors should be addressed:

- programmed funds,
- right of way availability and restrictions,
- environmental constraints, and
- Federal, State, and local regulations.

Funding will usually dictate the scope of a project and the options that are available. Oftentimes, a no pump alternative can provide a less costly project.

Right-of-way may be severely limited and the ability to acquire additional right-of-way may be minimal. Pump stations can often be designed for construction in limited space, but structural options exist, such as bridges and elevated highways on retaining walls, which can help minimize right-of-way requirements and avoid the need for pump stations.

If environmental concerns such as noise and visual impacts are primary factors, the alternatives may be limited. Political or public pressure may also limit alternatives. However, the designer should still look for the opportunity to avoid the need for a pump station.

4.6.3   Evaluation of Alignment Alternatives and Value Engineering

Though there is a difference between evaluation of alternatives and value engineering, they are essentially the same from the perspective of the drainage designer. The objective is to establish the most practicable and desirable drainage system. It is necessary to evaluate alignment options to determine:

- if a pump station can be avoided,
- if a pump station is the most economical alternative,
- if a pump station is the only feasible approach that meets design constraints/criteria, and
- how the alignment can be developed to minimize stormwater runoff to the pump station.

The intent of such evaluation is to balance economy with reliability. Generally, if a gravity system can be provided at a lower or similar life cycle cost compared to a pump station, the gravity system option is preferred. A gravity system is usually much more reliable than a pump station.

Detailed design procedures are not envisioned here, but an evaluation of the alternatives should be based on:

- roadway profile alternative s,
- highway vertical clearance requirements,
- access to adjacent businesses and property,
- outfall alternatives, and
- current usage of pump stations by highway agency in local area.

4.6.3.1   Roadway Profile Alternatives

Roadway profile alternatives include comparison of feasible options such as:

- elevated versus depressed sections,
- bridge structure and retaining wall for elevated sections versus retaining wall for depressed sections, and
- at-grade alternatives.

Some alternatives such as at-grade solutions will likely be limited by right-of-way restrictions. At-grade solutions may also be precluded by functional needs such as:

- multilevel directional interchanges,
- grade-separated interchanges, and
- railroad crossings.

4.6.3.2 Vertical Clearance

Vertical clearance requirements will depend on function and will affect the roadway profiles, which affects the comparison of options. For example, a highway underpass at a railroad will typically require lower clearance for the highway traffic than the railroad will require for an overpass. Thus the depressed section is likely to involve shorter vertical curves than the overpass option.

4.6.3.3 Access to Abutting Property

The alignment alternatives must consider access to abutting businesses and property. In some cases, an underpass and pump station combination may provide the least interference to driveway and street intersections for the same reason as discussed above. However, there are usually other alternatives such as providing local access roads and detours.

4.6.3.4 Outfall Alternatives

Even if a depressed section provides the most effective highway alternative, the designer should not automatically assume that a pump station is the only feasible drainage solution. The pump station is often considered necessary because the closest outfall is too high, or the tailwater is too high. There may be other locations within the watershed to which the depressed section could drain by gravity.

By evaluating the approximate potential life cycle costs of providing a gravity storm drain to an alternate outfall, and providing a pump station draining to the nearest outfall, the designer can determine which option is reasonable. The evaluation should not just be limited to construction costs. The construction cost of a gravity system could exceed those of a pump station, but a typical gravity system will have negligible to small operating and maintenance costs when compared to a pump station. No specific cost data is provided here because costs vary drastically by factors such as region, system type, construction method, and maintenance methods.

4.6.4 Roadway Profile Guidelines

Whenever a stormwater pump station is anticipated, the designer should develop the roadway profile based on:

- maximum practicable approach grades to the sag point,
- minimum practicable vertical curve lengths to sag point, and
- highest practicable sag point elevation.

These steps will minimize the roadway area that must drain to the sag point, thus help reduce the required pump and storage system sizes.

4.7 DESIGN PHASE

The preceding phases identify the need for, and establish the viability of, a stormwater pump station. The detailed design phase follows. This section describes the stages in the process, in which the detailed design of the pump station components is performed. Subsequent chapters provide the criteria and detailed procedures. The discussion is predicated on the assumption that the roadway geometry has been established.

4.7.1.1 Field Survey and Data Development

The data collection needs are similar to those required for normal drainage design of a highway project including establishing:

- utility sizes, locations, and ownership,
- outfall location, cross sections, profile grades, and roughness characteristics,
- right-of-way locations,
- roadway cross-sections, horizontal alignment, and vertical profile,
- drainage area sizes and runoff characteristics,
- collection system sizes, layout, and profile, and
- tie-in elevations.

4.7.1.2 Stage 1: Site Planning

The site planning stage involves evaluating suitable locations for the pump collection system, storage unit (if desired), and pump house. For detailed considerations, refer to Section 5.2 - Considerations for Site Planning.

4.7.1.3 Stage 2: Identification of Design Criteria

This stage involves establishing:

- design frequency,
- peak outflow (if appropriate),
- station type (wet-pit station or dry-pit station),
- design philosophy,
- minimum storage volume,
- maximum allowable highwater, and
- discharge velocity.

The criteria must be established prior to proceeding with design of the storage system and pump system. Refer to Chapter 6 for discussion on these and other considerations.

4.7.1.4 Stage 3: Hydrologic Analysis

The hydrology stage involves determining:

- drainage boundary and area,
- runoff characteristics,
- design runoff hydrograph(s), and
- cumulative inflow.

The designer may begin hydrologic analysis after establishing the roadway alignment and design criteria. State of the practice hydrologic methods are employed such as appear in the FHWA publication Highway Hydrology.[5]

4.7.1.5 Stage 4: Determination of Storage

This stage comprises designing or evaluating the following for providing storage volume:

- collection system,
- storage unit, and
- wet well.

4.7.1.6 Stage 4a: Collection System Design/Evaluation

The collection system stage establishes the collection system layout, element sizes and shapes, and evaluates the storage in the collection system between a maximum allowable high water and a minimum elevation in the pump station. Some agencies specifically design parts of the collection system by iteration to provide storage, in lieu of a bona fide storage unit, to minimize the required pump capacity.

4.7.1.7 Stage 4b: Storage Unit Sizing

Many highway agencies design a specific storage unit to meet minimum desired storage volumes or to optimize the required storage volume and pumping capacity. If such a unit provides only the minimum required volume, the sizing process can begin after the roadway alignment and design criteria have been established. If optimal storage is to be provided, the process may proceed as indicated in the flowchart, but several iterations of storage unit sizing and pump configuration may be necessary.

4.7.1.8 Stage 4c: Pit Dimensions and Wet Well Storage

One of two approaches is usually taken:

- designer chooses a preliminary pit type and size, determines what pumps are needed, and then ensures clearances meet appropriate specifications for needed pumps, or
- designer ignores wet well storage and establishes pump sizes and number, then sizes the pit based on appropriate criteria.

The approach is usually a matter of choice. Either approach will often result in some level of iteration because the pump capacity is dependent on total storage (of which wet-well storage is a component), and wet-well storage is dependent on pit dimensions. Minimum pit dimensions are dependent on pump number and capacity. When storage units are used, the wet well storage is not usually the major part of system storage and minor size adjustments may not alter the pump selection.

4.7.1.9 Stage 5: Trial Pumping Configuration

The design of a typical pump station is an iterative process. The only condition under which a closed solution is likely is the combination of a triangular/trapezoidal runoff hydrograph method and one pump. (A minimum of two pumps is recommended for all highway pump stations.)

The process requires the designer to establish trial types, sizes and numbers of pumps. The subsequent stages will evaluate the performance of the trial configuration. If the configuration proves to be either oversized or undersized, the designer must establish a new configuration and repeat the evaluation.

The trial pump configuration may follow establishment of the design criteria. The following table shows the necessary order of the design process depending on the selected or required approach.

If the design approach is to:	Then:
Fix the peak outflow	The trial pump configuration may be established after development of cumulative inflow volume (mass inflow curve) and before evaluating storage.
Optimize storage/pump capacity	The trial pump configuration should follow hydrograph development and trial storage system evaluation. A series of iterations of storage size and pump configuration will be necessary to establish which combination achieves the best balance of pump size, storage unit size, and construction cost
Fix the storage volume	The trial pump configuration should follow development of cumulative inflow volume and trial storage system evaluation.

4.7.1.10 Stage 6: Trial System Evaluation

The trial system evaluation stage includes various procedures to determine whether or not the trial pump configuration and storage system achieve the following objectives:

- peak outflow equal to or lower than target peak (if appropriate),
- highest water level in station does not exceed maximum allowable high water for design storm,
- pump configuration is not excessive, and
- selected pumps comply with manufacturer's specifications, such as cycling time and Net Positive Suction Head (NPSH).

The evaluation includes the following procedures, a suggested order of which appears in the Design Process Flowchart:

- development of inflow hydrograph,
- development of mass curve from inflow hydrograph,
- establishing trial storage unit size,
- selection of initial wet well dimensions,
- development of stage versus storage relationship,
- establishment of usable storage,
- establishment of trial pump switching/sequencing (cut-on, cut-off elevations),
- mass curve routing,
- routing sufficiency checks,
- determination of Total Dynamic Head,

- development of system head curves,
- establishing an operation range,
- establishing power requirements,
- selection of pumps and piping from manufacturers' performance curves and catalogue data,
- performing cycle time checks, and
- design optimization (balancing storage with pump type and size), if desired.

4.8   DEVELOPMENT OF PLANS, SPECIFICATIONS AND ESTIMATES

This phase (PS&E) involves developing the details, specifications, and quantity estimates for the project.

4.8.1   Layouts and Details

Site-specific layouts, special details, and standard details should include the following:

- general site layout,
- collection system plan, profile, and details,
- storage unit plan, elevation, and details
- well plan, elevation, and details,
- pump house plan, elevation, and details,
- discharge piping and outfall plan, profile, and details, and
- electrical and mechanical plans and details.

4.8.2   Specifications, Provisions and General Notes

The designer should used standard specifications to the maximum degree practicable. However, pump stations will usually require special specifications, special provisions, and general notes to include items such as:

- pump performance specifications and tolerances,
- pump installation and testing, and
- special construction requirements.

Additionally, it is recommended practice to establish an operation and procedures manual during PS&E that is to be used after construction of the pump station.

4.8.3   Design Review

The highway agency should provide an internal review of the design to ensure:

- adequacy of assumptions and criteria,
- accuracy of calculations and details,
- compliance with design criteria, policy and regulations, and
- conformance with any commitments made in environmental documents (including Environmental Impact Statement or Environmental Assessment).

4.9 CONSTRUCTION AND TESTING

This phase, though typically beyond the scope of the designer's responsibilities, must be considered by the designer. Close coordination is required between construction personnel and designers to ensure that the design makes appropriate accommodation for construction needs and that any lessons learned from construction can be transferred to the designers.

Pump performance tests are conducted in place to ensure that pumps perform as specified. Periodic tests during the operational life of the station are appropriate to check continued operating efficiency of the pumping station.

4.10 OPERATION AND MAINTENANCE PHASE

Operation and maintenance of pump stations involves frequent inspection, monitoring, and maintenance. The highway agency should establish operation and maintenance procedures and schedules. Some agencies perform their own inspection and maintenance, while others pass on these responsibilities to local agencies or independent contractors.

4.11 RETROFITTING STAGE

The retrofit stage is inevitable for all stations that ultimately are not completely replaced. Retrofit involves replacing various components of the pump station, usually the pumps themselves. Hopefully this stage is only necessary because the components of the station have passed their design life expectancy. Refer to Chapter 13 – Retrofitting Existing Systems for a discussion of the need for retrofitting and appropriate measures to take.

This page intentionally left blank.

5. SITE PLANNING AND HYDROLOGY

5.1 INTRODUCTION

This chapter discusses considerations for establishing suitable locations for pump stations, identifies reference material for hydrologic methods and presents some hydrologic criteria for design.

5.2 CONSIDERATIONS FOR SITE PLANNING

5.2.1 Drainage Area

A primary consideration for the design of a stormwater pump station is the drainage area. Pump station capacity requirements and associated costs increase with increasing drainage area. Therefore, the designer's goal should be to minimize the area that will drain to the proposed pump station to the degree practicable.

The primary means of minimizing the drainage area are:

- maximize roadway grades and minimize vertical curve lengths in the approach to the roadway low point(s) to reduce the length of roadway that drains to the low point(s),
- use gravity storm drains as deep as practicable to drain as much surface area as practicable,
- use retaining walls, where practicable, to minimize width of depressed roadway section, and
- prevent offsite runoff from flowing to pump station using features such as berms and interceptor swales.

This approach is somewhat contrary to the typical goals of roadway and storm drain design. Usually, flatter roadway grades are preferable to steep roadway grades and shallow storm drains are preferred to deep storm drains. The pump station designer must coordinate with the roadway designer to ensure that the roadway design criteria are met.

Figure 5-1 shows a railroad underpass in Houston, Texas that required a pump station. Roadway profile grades were kept as steep as practicable for the design speed in order to help reduce the drainage area. The retaining wall, which was used because of limited right-of-way, also helped reduce the drainage area. In addition, storm drain inlets were used in the upper ends of the sag curve to direct as much flow as possible away from the pump station collection system.

Figure 5-1. Railroad underpass

5.2.2   Proximity

The low point (or sag) for a highway stormwater pump station, is usually at an underpass such as at a highway intersection or railroad crossing. Generally, it is neither practicable nor preferable to place a pumping station immediately adjacent to the low point of the highway because of the following factors:

- the station would usually be susceptible to flooding in the case of malfunction,
- there would likely be an inadequate safety buffer for maintenance personnel,
- the housing could pose a hazard to traffic,
- potentially high construction costs, and
- access to the station would likely be difficult or require interruption of traffic.

The station floor, on which the electrical equipment and other facilities are placed, should be set at a higher elevation and the housing situated away from the highway lanes. In the case of small stations with submersible pumps, the wet well can reasonably be at the low point, provided the electrical control panel is located on higher ground. Suitable locations for the station include:

- outside a frontage or service road,
- between the highway main lanes and the frontage road,
- at natural grade at the top of the cut slope of an underpass, and
- on the berm within the easement of an outfall ditch.

The location of all features associated with the pump station should conform to highway clear zone requirements.

The pump station shown in Figure 5-2 is set above and away from the roadway lowpoint.

Figure 5-2. Pump station for a depressed section

5.2.3   Site Access

Pump stations require frequent inspection and maintenance. Therefore, provisions should be made for easy access to the station and so that the station is compatible with the number and size of vehicles and hoisting equipment that will likely be required to construct and maintain the station. Such provisions should include:

- service road/driveway with suitable turning radii. See AASHTO's *A Policy on Geometric Design of Highways and Streets* for guidance,[7]
- off-street parking,
- station loading area,
- turn around area,
- space for heavy lifting equipment, and
- roadside warning signs.

The site shown in Figure 5-3 has poor access because there is no vehicular access from ground level. Steps provide access from the highway shoulder as can be seen in Figure 5-2. It will be difficult to replace any equipment, and traffic will be interrupted during any significant maintenance.

Figure 5-3. Limited access to pump station from shoulder of highway

The site shown in Figure 5-4 provides safe access via a driveway, without interfering with traffic. The depressed freeway is behind this facility.

Figure 5-4. Driveway access to pump station in Arizona

5.2.4 Power Sources

The designer should consider how power will be supplied. There will usually be the need to supply electricity and sometimes natural gas. The designer will need to communicate with appropriate power suppliers to locate the site within easy reach of the power supply. If all power is to be generated on site, then location concerns will relate primarily to placing the site within easy access of a roadway for fuel supply.

5.2.5 Storage

When considering the location of a highway stormwater pump station, it is necessary to ensure that there is enough space between the highway low point and the pump house to provide storage. Storage is an essential component of a highway stormwater pump station because it allows the designer to reduce the required total pumping capacity. A minimum storage capacity is required to ensure that pump cycling criteria are met. See Chapter 6 - STORAGE SYSTEM for detailed considerations of storage components and the effect of storage. See Section 7.2.6 - Pump Cycling for the effect of storage on cycling.

5.2.6 Environmental Quality

During the planning phase, an Environmental Assessment (EA) or Environmental Impact Statement (EIS) may be necessary. Usually, the proposed pump station is incorporated into the study of a whole stretch of highway improvement. Any commitments that are made in the EA or EIS should be incorporated into the design.

Primary environmental quality issues to be considered include:

- visual impact,
- air quality,
- noise attenuation, and
- water quality.

5.2.6.1 Visual Impact

Most state transportation agencies now have on-staff or consultant architects and landscape architects. It is preferable for the designer to coordinate with such individuals to identify cost-effective ways of enhancing visual quality. Typical low-cost measures include:

- providing cut and embankment slopes of 1:3 or flatter,
- allowing maximum areas of natural or planted vegetation,
- enclosing unsightly objects such as storage tanks,
- using submersible pumps to reduce the size of required above-ground facilities,
- using local building materials that blend in with the surrounding architecture, and
- providing underground utilities (power supply, phone lines, etc.).

5.2.6.2 Air Quality

To minimize potential air quality issues, where electrical generators or engine-driven pumps are used, LPG or natural gas are preferred to gasoline and diesel because they have cleaner byproducts (water vapor and carbon dioxide). At the pump station site, the least impact results from a purely grid-supplied electrical system.

5.2.6.3 Noise

Noise attenuation is often a primary concern near residential areas. Where practicable, it may be desirable to minimize the potential for noise pollution by taking one or more of the following measures:

- use submersible pumps
- where submersible pumps are not practicable, use electrically-driven motors
- if engines are used, provide appropriate mufflers
- build the pump house out of concrete or masonry

Sound insulation of the pump house walls may be an option; however, the effectiveness of such a measure may be reduced by necessary station ventilation.

5.2.6.4 Water Quality

Water quality issues have become more prevalent and require attention for pump station design. For surface water, Environmental Protection Agency (EPA) and state regulations will likely indicate the need for some level of best management practices (BMPs) to help reduce the potential for polluting receiving waters. The primary regulations are the National Pollution Discharge Elimination System (NPDES) requirements for construction (Industrial Activities) and NPDES Municipal Separate Storm Sewer System (MSSSS) permit provisions. A highway stormwater pump discharge is considered a non-point source just like a storm drain outfall. Therefore, if the discharge point is within a municipality that is currently operating under an MSSSS permit, the designer must ensure that BMPs conform to the provisions of the permit by providing appropriate features outside the pump station. The permit holder and status will vary by state and municipality and the number of municipalities covered by the NPDES requirements is subject to change.

5.2.7 Safety

Safety must be a primary consideration for all pump station design and should include provisions for:

- construction personnel,
- inspection and maintenance personnel,
- motorists, and
- the general public.

Provision for adequate access is a primary safety measure for inspection and maintenance personnel. Other considerations include meeting OSHA requirements for station access holes, hoisting, steps, ventilation etc. Refer to Chapter 12 – Construction, Operation and Maintenance for additional discussion of inspection and maintenance considerations.

Primary means of ensuring public safety include:

- minimizing traffic hazard by suitable site location (See Section 5.2.2 - Proximity),
- providing warning signs,
- meeting clear zone requirements for the highway or providing appropriate protection,

- providing adequate security (See Section 5.2.8 - Security), and
- providing failure and high water level alarms.

5.2.8 Security

A pump station is often an attraction for children and vandals. The site should be protected both during and after construction. The primary security measures are:

- perimeter fencing,
- intruder alarms,
- concrete or masonry housing, and
- locked louvered windows.

5.2.9 Utilities

In addition to a power supply the following utilities are either essential or preferred:

- water supply,
- sanitary sewer,
- local storm drainage, and
- phone line.

A potable water supply may be desirable to:

- flush the sump,
- facilitate cleaning of the station,
- irrigate the station grounds, and
- accommodate toilets and wash basins, if provided in the pump station.

5.2.10 Soil Investigations

Consideration of local soil conditions and characteristics may affect the proposed location of a pump station. The hydraulic designer should coordinate the site planning with a geotechnical engineer to ensure that the proposed location is not likely to encounter significant problems associated with water table levels, soil bearing capacities, plasticity, and seismic activity. For example, clay with water-bearing, sand lenses can pose problems when constructing the pump sump, especially for sunk caissons which can experience differential settlement. The geotechnical engineer also may suggest construction methods that are appropriate for the soil conditions.

5.3 HYDROLOGIC CONSIDERATIONS AND CRITERIA

5.3.1 Premise for Design

A highway stormwater pump system is designed to remove runoff from the highway system without inundating the highway under design conditions. Usually the drainage collection points are in depressed sections so that there is no relief location to which the runoff may drain.

Therefore, a typical highway stormwater pump station should be designed to accommodate not just frequent runoff events but also less frequent events.

5.3.2   Design Frequency

The recommended design for a highway stormwater pump station is a return period of 50 years with the following qualifications:

- mandatory for interstate highways,
- may be reduced or increased if supported by risk assessment,
- may not be practicable for retrofit pumps, and
- where the pump is used to isolate the highway from the backwater effects of a receiving water, the designer should consider the implications of joint probabilities of flooding on the highway system and the receiving water.

5.3.3   Check Frequency

The standard check flood is one having a return period of 100 years. The primary intent of this check is to identify the potential risks associated with inundation during such an extreme event and ensuring conformance to Federal Emergency Management Agency (FEMA) National Flood Insurance Program (NFIP) criteria.

In addition, it is recommended that a frequent event, such as a 2-year, be used to check low flow operation so as to ensure that no excessive cycling is likely.

5.4   DESIGN HYDROGRAPH

The design runoff hydrograph is the basis for sizing a highway stormwater pump system. Figure 5-5 shows a typical design inflow hydrograph. The method used should be applicable for the conditions to be addressed and based on the design frequencies discussed above. Refer to the FHWA publication "Highway Hydrology", HDS 2,[5] for detailed considerations and procedures for determining runoff hydrographs.

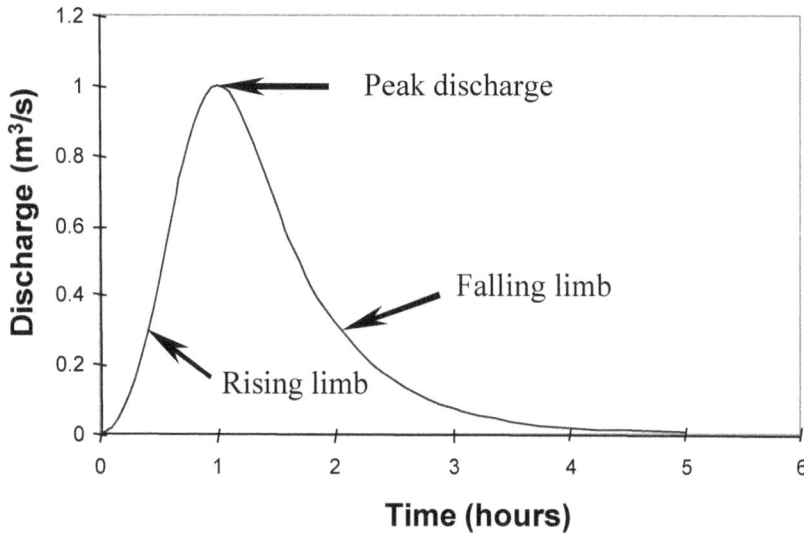

Figure 5-5. Typical runoff hydrograph

5.4.1 Hydrologic Methods

The choice of hydrologic methods is either based on agency standard practice or left up to the designer based on local conditions and judgment. Specific methods often used by state highway agencies include:

- National Resources Conservation Service (NRCS) – formerly Soil Conservation Service) – Tabular Hydrograph Method of *Technical Release 55, Urban Hydrology for Small Watersheds* (TR55),[8]
- NRCS dimensionless unit hydrograph method based on *Technical Release 20, Hydrology*, (TR 20),[9] and
- Methods available in the computer programs HEC-1 and HEC-HMS.

5.4.1.1 NRCS Methods

The TR 55 tabular approach is simple and often sufficient. With the availability of powerful computers, spreadsheets, and hydrologic software, hydrographs from methods such as the TR 20 method are simple and quick enough to generate and are more accurate (TR 55 is a simplification of TR 20). However, when using TR 20, the designer should consider the validity of using long duration rainfall events, such as 24 hours. Shorter duration storms that compare with the estimated time of concentration for the drainage area are usually more appropriate for pump station design. The TR 20 method is applicable to shorter duration events if an appropriate design rainfall distribution is available. Standard rainfall hyetographs for the country are limited to 6-hour and 24-hour durations.

5.4.1.2 Other Methods

Other local methods may be used when the agency is comfortable that the discharge versus frequency relationships are statistically based and the method allows the designer to choose an appropriate design rainfall duration.

5.5 PROCEDURE TO DETERMINE MASS INFLOW

A mass inflow curve represents the cumulative inflow volume with respect to time. In order to determine a mass inflow curve, the designer first must have developed an inflow hydrograph. The following describes how to develop a mass inflow curve.

Step 1.   Evaluate the time base of the design hydrograph and select a time increment. Usually, the designer uses the same time increment as that used for developing the inflow hydrograph.

Step 2.   Develop a table of time, time increment, inflow rate, average inflow rate, incremental inflow volume and cumulative inflow volume as shown in Table 5-1.

Step 3.   At each time step, take the inflow rate from the computed inflow hydrograph. See Column 3 in Table 5-1.

Step 4.   Compute and tabulate the average inflow rate as half of the sum of the current and previous inflow rates for each time step. See Column 4 in Table 5-1.

Example: using Table 5-1, the average inflow at time = 5 mins is:
$(0 + 0.003) / 2 = 0.0015$ m^3/s (0.053 cfs)

Step 5.   Compute and tabulate the incremental volume for each time step as the average inflow rate multiplied by the time increment (in seconds). See Column 5 in Table 5-1.

Example: using Table 5-1, the incremental inflow volume at time = 5 mins is:
0.0015 m^3/s x 5 mins x 60 secs/min = 0.45 m^3 (1.48 ft^3)

Step 6.   Compute and tabulate the cumulative volume for each time step as the sum of the incremental volume in the current time step and the cumulative volume of the previous time step. See Column 6 in Table 5-1.

Example: using Table 5-1, the cumulative inflow volume at time = 5 mins is:
$0 + 0.45 = 0.45$ m^3 (1.48 ft^3)

Step 7.   Plot a curve of cumulative volume versus time. The result is a mass inflow curve.

Example: Refer to Figure 5-6 which is based on Table 5-1.

Table 5-1. Development of cumulative inflow

1	2	3	4	5	6
Time	Time Increment	Inflow Rate	Average Inflow	Incremental Inflow	Cumulative Inflow
(Mins)	(Mins)	(m^3/s)	(m^3/s)	(m^3)	(m^3)
0		0		0.0	0.0
5	5	0.003	0.0015	0.45	0.45
10	5	0.006	0.0045	1.35	1.80
15	5	0.009	0.0075	2.25	4.05
20	5	0.011	0.0100	3.00	7.05
25	5	0.014	0.0125	3.75	10.80
30	5	0.017	0.0155	4.65	15.45
35	5	0.02	0.019	5.55	21.00
40	5	0.023	0.022	6.45	27.45
45	5	0.025	0.024	7.20	34.65
50	5	0.028	0.027	7.95	42.60
55	5	0.031	0.030	8.85	51.45
60	5	0.034	0.033	9.75	61.20
65	5	0.071	0.053	15.75	76.95
70	5	0.127	0.099	29.70	106.7
75	5	0.326	0.227	67.95	174.6
80	5	0.538	0.432	129.6	304.2
85	5	0.609	0.574	172.1	476.3
90	5	0.481	0.545	163.5	639.8
95	5	0.34	0.411	123.2	762.9
100	5	0.184	0.262	78.6	841.5
105	5	0.142	0.163	48.9	890.4
110	5	0.113	0.128	38.3	928.7
115	5	0.099	0.106	31.8	960.5
120	5	0.093	0.096	28.8	989.3
125	5	0.076	0.085	25.4	1014.6
130	5	0.071	0.074	22.1	1036.7
135	5	0.065	0.068	20.4	1057.1
140	5	0.059	0.062	18.6	1075.7
145	5	0.057	0.058	17.4	1093.1
150	5	0.054	0.056	16.7	1109.7

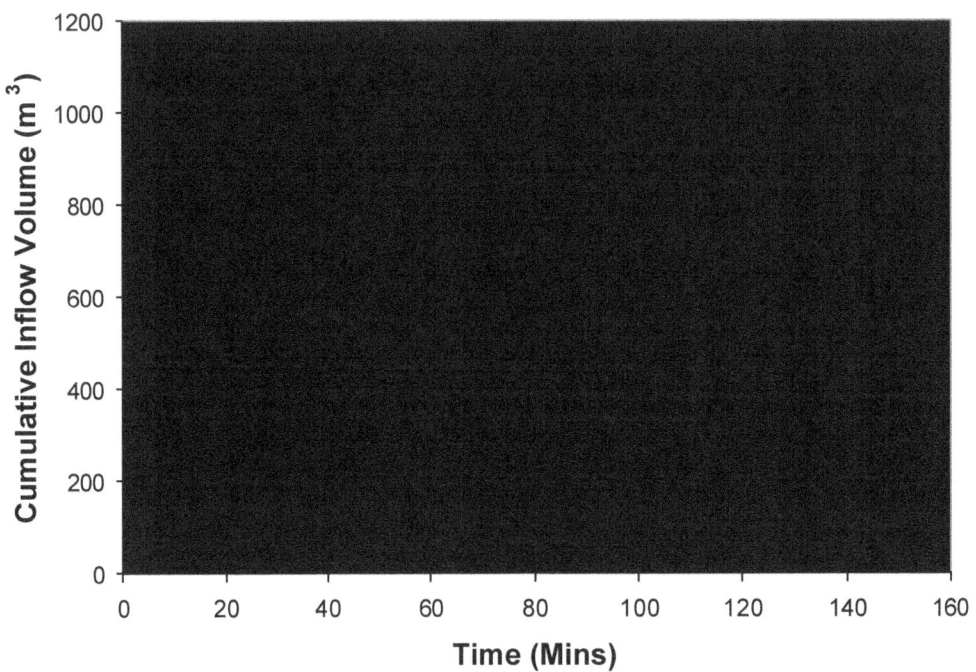

Figure 5-6. Typical mass inflow curve

6. STORAGE SYSTEM

6.1 INTRODUCTION

The biggest distinction between highway stormwater pump stations and other pump systems is the significant use of storage. This section:

- identifies the components of the system that provide storage,
- indicates criteria used to establish storage, and
- provides guidance on sizing storage elements.

6.2 STORAGE COMPONENTS

The storage components of a highway stormwater pump station include:

- the collection system,
- the storage unit, and
- the wet well system.

6.3 STORAGE CONCEPTS

6.3.1 Storage Potential

The design of stormwater pump stations is analogous to the concept of stormwater detention. In other words, the rate of outflow is the difference between the rate of inflow and the rate of change of volume stored. For typical detention ponds, as the storage volume increases, the peak outflow rate decreases. For a pump station, as the storage volume increases, the required pumping capacity decreases. Figure 6-1 shows a typical inflow hydrograph routed through a pump station. The area bounded by the rising limbs of the inflow and outflow hydrographs represents the total volume in storage.

For a particular design inflow hydrograph, the peak pumping capacity requirements reduce with increasing storage. At one extreme, if no storage were available, the pump system would be required to discharge at the same rate as the inflow, requiring a large number of varied-capacity or variable speed pumps and expensive, complex control systems. At the other extreme, if the storage exceeded the inflow volume, only a low flow pump would be required to evacuate the system, but the cost of storage would likely be excessive. Somewhere in between, there is an optimal condition for a particular site that balances the storage provided with the size and number of pumps.

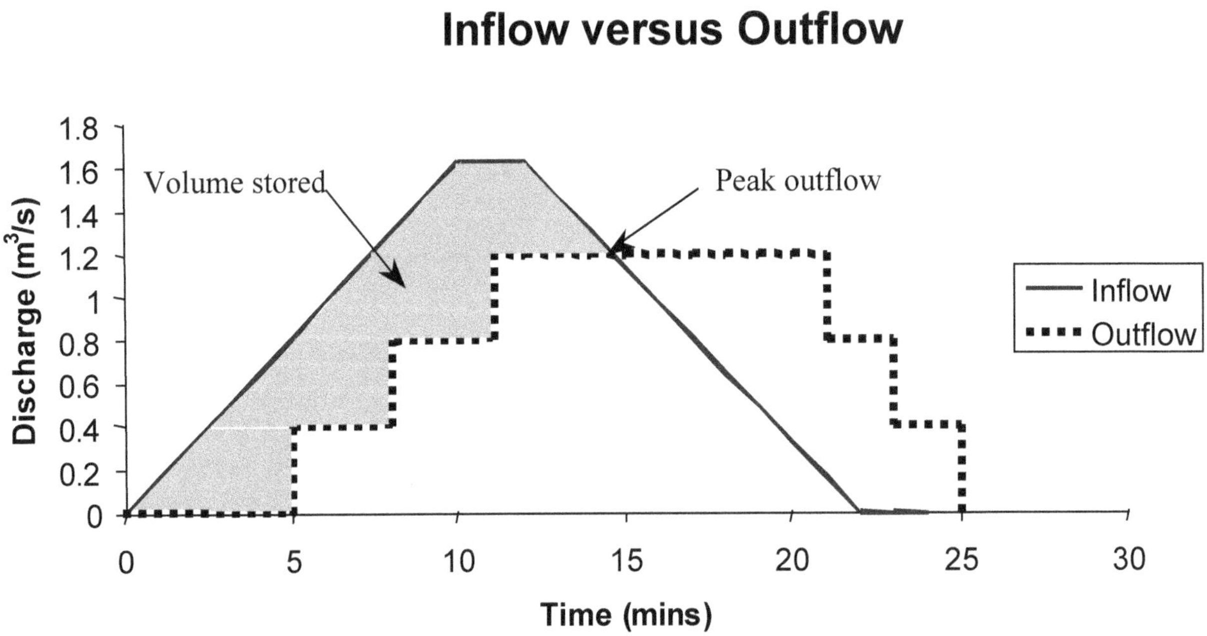

Inflow versus Outflow

Figure 6-1. Hydrograph routed through pump station

6.3.2 Storage Design Philosophies

There are three general approaches to pump system storage and pump sizing.

- Size the collection system for adequate conveyance of the runoff using conventional storm drain design procedures (See FHWA's *Urban Drainage Design Manual*[10]). Choose a wet well size, size the pumps and set the pump switching to maximize use of the provided storage.
- Identify a target outflow rate and choose pumps that have a total capacity of the target outflow rate. Establish the required system storage to detain the design inflow hydrograph. This approach may be used in areas where ordinances limit peak discharges to pre-developed rates, or where the outfall system has limited capacity.
- Using iteration, vary the storage volume, pump sizes and number of pumps to determine a combination that minimizes total life-cycle cost. This is a more time-consuming approach but will most often produce the best overall design.

The latter approach is preferred since neither of the other two approaches identify the combination of pump size and storage that minimizes life-cycle costs.

The primary effect of storage is to attenuate peak discharge rates. Additionally, storage affects the frequency of pump operation (pump cycling times). As storage volume increases, the cycling time increases. Longer cycling times are preferred and design cycling times must exceed manufacturer's specified minimum cycling times. Therefore, it will be necessary to ensure that either enough storage is provided for a specific system of pumps, or a system of pumps and sequencing scheme is developed to accommodate the storage.

6.3.3 Available Storage

The total storage volume in the collection system, storage unit, and wet well between the lowest pump operation elevation and the maximum allowable high water serves to detain the inflow. This total storage volume is termed total available storage in this document.

At any stage, H, the volume used for storage of the inflow is the volume between the water level and the lowest pump operation elevation (Stage = 0) as shown in Figure 6-2. The available storage is independent of inflow rate if the hydraulic grade line is ignored. See Column 5 in Table 5-1

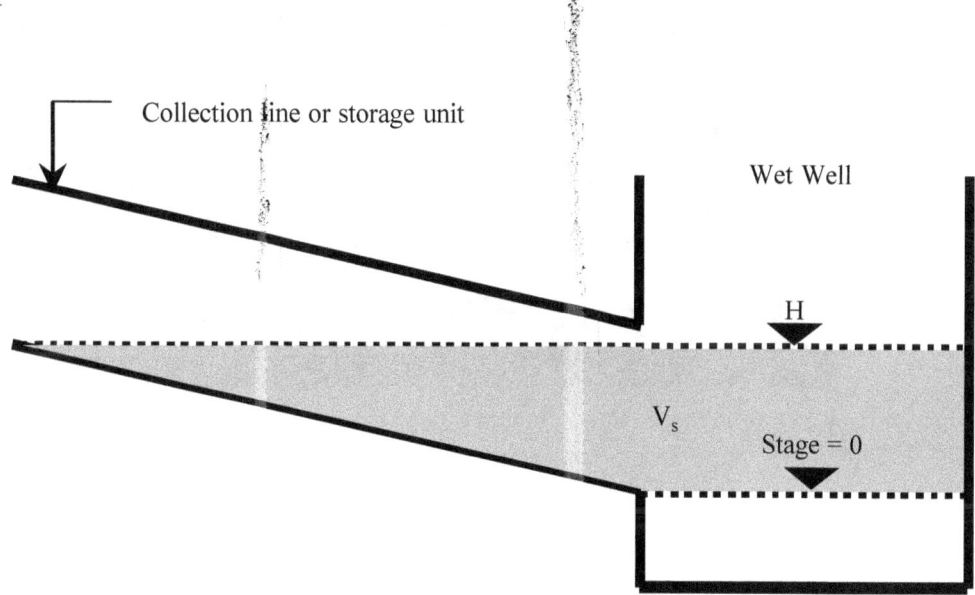

Figure 6-2. Volume in storage at any stage

6.3.4 Procedure to Estimate Total Available Storage Required

Step 1.   Select a design inflow hydrograph

Step 2.   Choose trial target total pumping rate.

             Example: choose 0.4 m³/s (14 cfs).

Step 3.   Using a plot of the inflow hydrograph, draw an estimated outflow hydrograph (based on the target pumping rate) from a tangent on the lower portion of rising limb to the falling limb at a flow equal to the target total pumping rate.

             Example: refer to Figure 6-3.

Step 4.   The area between the estimated pumping rate line and hydrograph represents the estimated required volume (V_{req}).

Example: In Figure 6-3, the volume is estimated to be 260 m^3. This is determined by breaking up the area into approximate right triangles and calculating the area of the triangles.

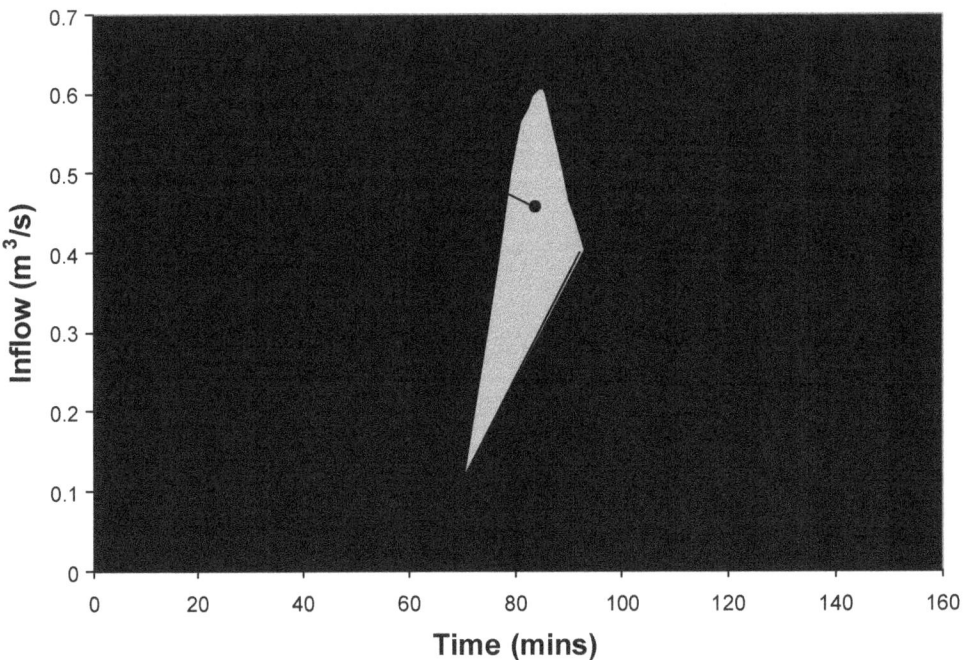

Figure 6-3. Estimate of storage required

6.4   COLLECTION SYSTEM

A collection system conveys stormwater runoff from the highway to the pumping station. A typical collection system includes some combination of the following elements:

- ditches,
- roadway gutters,
- inlets,
- access holes,
- channels,
- conduits, and
- trash gates or racks.

For detailed discussion on planning and design of these elements refer to publications such as Hydraulic Engineering Circular No. 22, Urban Drainage Design manual,[10] and Hydraulic Design Series No. 4, Introduction to Highway Hydraulics.[11]

Generally, the roadway gutters, while part of the collection system, are not available for consideration as storage volume because they are higher than the allowable high water, by definition. Only those elements that provide volume between the maximum high water elevation and the higher of the sump system inlet flow line elevation and the lowest pumping elevation will affect the design of the pump system.

Elements of the collection system should be laid out to alignments, elevations and slopes to suit the highway. No matter which design philosophy prevails (See Section 6.3.2 - Storage Design Philosophies), the maximum storage capacity can be achieved by keeping as much of the collection system with a flow line elevation as close to the sump inlet elevation as possible. The flow line of the collection system should be kept as high as possible to minimize the depth of the wet well inlet invert. However, the collection system must maintain a steep enough slope to reduce the potential for sedimentation.

6.4.1 Procedure to Establish Collection System

Step 1.   If not already done, design the collection system in accordance with the standard storm drain design principles and the pertinent considerations and criteria that appear in Chapter 7. Refer to the FHWA publication, *Urban Drainage Design Manual, HEC-22*,[10] for guidance on storm drain design.

Step 2.   Prepare a simple layout and profile of the proposed collection system, including the storage unit if applicable, with annotation that includes:

- conduit section dimensions,
- conduit lengths,
- conduit invert elevations at nodes, and
- inlet invert and edge of pavement elevations at inlet openings.

6.5   MAXIMUM HIGHWATER ELEVATION

The maximum highwater in the sump is the elevation that does not cause the hydraulic grade line for the design event in the collection system to exceed the allowable high water at the roadway low point.

6.5.1   Roadway Low Point

An allowable high water should be set some depth below the throat of the lowest inlet in the collection system. The throat of the inlet refers to the point in the inlet opening that controls the flow entering the inlet. Usually, the lowest inlet will be at the sag point of the roadway. Preferably, the depth of the allowable highwater below the throat of inlet should be 0.6 m (2 ft) or more to provide a safety margin. However, in flat terrain, in may not be practicable to attain such a safety margin.

6.5.2   Hydraulic Grade Line

The hydraulic grade line should be computed based on the design peak inflow rates through the collection system. Refer to the FHWA publication *Urban Drainage Design Manual, HEC-22*,[10] for the procedure.

6.5.3   Procedure to Determine Allowable Highwater in Wet Well

Step 1.   Establish the maximum allowable highwater at the roadway low point ($HW_{R(max)}$) as the roadway elevation at the lowest inlet opening less a safety factor of 0.3 m – 0.6 m (1 ft – 2 ft).

Step 2.   Compute the hydraulic grade line from the wet well to the lowest inlet in the roadway using the following:

- annotated layout and profile of the collection system and wet well,
- computed design peak inflow rates,
- trial water level in wet well. The first trial level should be below the maximum allowable highwater at the roadway low point ($HW_{R(max)}$) but above the soffit of the wet well inlet conduit, and
- procedures presented in the FHWA publication, *Urban Drainage Design Manual*.[10]

Step 3.   Compare the hydraulic grade line (HGL) at the lowpoint of the roadway with the maximum allowable highwater at the roadway low point ($HW_{R(max)}$). Use the following decision table:

If HGL is:	Then:	Comment
Higher than $HW_{R(max)}$	Repeat Step 2 using a lower trial water level in wet well.	The trial level exceeds allowable level
More than 75 mm (3 in.) lower than $HW_{R(max)}$	Repeat Step 2 using a higher trial water level in wet well.	The trial level is too low
At or just below $HW_{R(max)}$	Proceed to Step 4.	The trial level represents $HW_{s(max)}$.

Step 4.   Compare the resulting $HW_{s(max)}$ to the soffit elevation of the storage unit at the intake. If $HW_{s(max)}$ is below the soffit, then the storage unit will not be completely full. The options for proceeding are:

- Reduce the flowline of the collection system, adjust the profile plot and annotation, and return to Step 1. Note that this will result in larger pump power requirements.
- Reduce the height of the conduit system components and increase the width or number of barrels to maintain design capacities. Adjust the profile plot and annotation, and return to Step 1.
- Proceed without adjusting the storage system. This is not preferred because of wasted volume.

6.6   WET WELL SYSTEM

From a hydraulic design perspective, consideration of the wet well at this stage is limited to the contribution of volume to the total available volume and the available volume at increasing water levels. A local or preferred standard dimension and shape is usually chosen and used for initial storage calculations or the storage in the wet well is ignored for the trial design. The final dimensions of the sump are established after specific pumps have been selected. Refer to Section 9.1 - Sump Size and Clearance Criteria for sump dimension criteria.

6.6.1 Procedure to Select Preliminary Wet Well Dimensions

Minimum pump pit dimensions depend on the number, size and type of pump, but the pump size depends on the storage provided, a part of which is provided by the wet well size. In some cases, the wet well may provide significant storage, and other case may provide only nominal storage. Here, the overall wet pit dimensions are chosen as a first estimate. Detailed sump size and clearance checks are performed after pump sizing and selection.

Step 1.   Choose a trial wet well size. Wet well sizes and shapes will vary according to highway agency preferences and type of station. For typical well types, refer to Section 3. - Pumps and Pump Station Types. At this stage, only the overall internal dimensions and shape are necessary.

Step 2.   Establish the lowest pumping elevation. For some typical details, this will be the wet well inlet invert elevation. For others, the lowest pumping elevation will be some preset depth below the wet well inlet invert elevation depending on the wet well details.

Step 3.   Add the wet well and dimensions to the collection system layout and profile. The bottom elevation of the sump is not needed at this stage.

6.7 STORAGE UNIT

A storage unit is provided to achieve a desired storage volume in excess of the storage provided by the collection system. A storage unit is an integral component of the dry-pit station, where it also serves as the wet well. It is not required for a wet-pit station; however, it is often used to achieve a desired storage volume when the designer wishes to reduce the required pump capacity. The storage unit can also serve as a fire protection unit and as a sand trap.

The storage unit can be made of various materials and shapes, but concrete boxes have been predominant. A typical storage unit is illustrated in Figure 6-4.

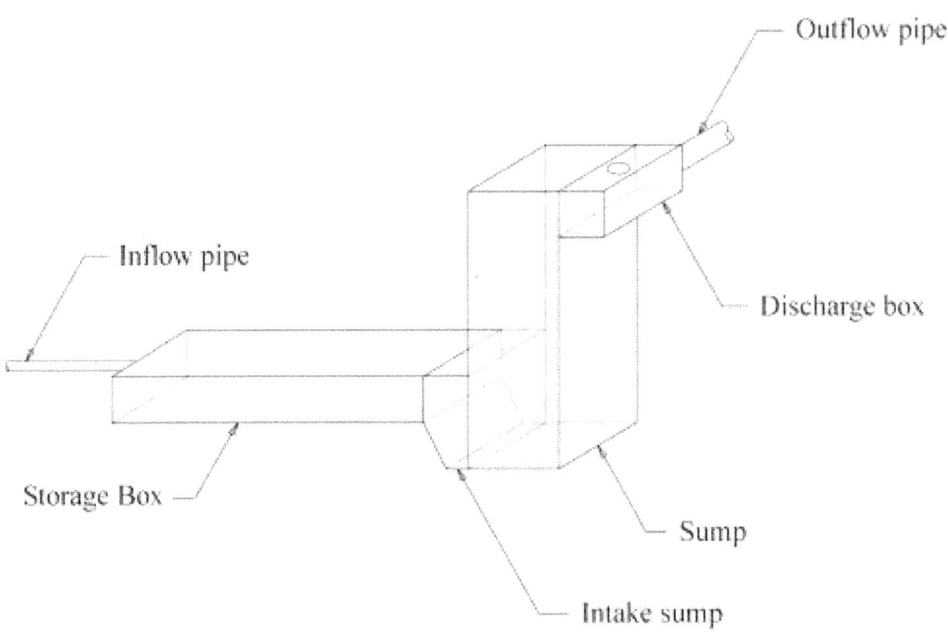

Figure 6-4. Typical storage unit

To minimize right-of-way requirements, single or multiple boxes or pipes are placed parallel to the highway or transversely under the highway at the low point of the collection system and depressed area. The boxes are sized to store as much of the peak flow volume from the design storm as is economically feasible. The storage unit may be designed to accommodate temporary power outages or pump failures; however, the provision of backup power and redundancy in the pumping capacity is preferred.

A minimum storage volume can be established that will allow a hazardous highway spill to be contained in the storage unit and its sand trap until pumped out. In the event of a spill, the pump station would not be contaminated or damaged.

The storage unit should have adequate access to allow for inspection and maintenance. A frequent maintenance item will be the need to remove sediment deposits and trash, so the type of access should be compatible with the anticipated clean out method. For single barrel systems, access hole spacing should conform to state policy. For multi-barrel systems, access to each barrel should be provided either by individual access holes or by horizontal access windows between the barrel walls.

6.8   STORAGE CALCULATIONS

Table 6-1 provides some quick reference equations for common sections used in wet wells and collection systems.

The variables are defined below:

d =   deepest depth in element, m (ft)

s =   slope of element, m/m (ft/ft)

s_s = side slope (ratio of vertical to horizontal), m/m (ft/ft)

D = diameter or rise, m (ft)

B = breadth (horizontal plane), m (ft)

L = length of element, m (ft)

W = width of element, m (ft)

V = volume of element, m³ (ft³)

The FHWA software program HY-22[16] can be used in conjunction with HEC-22[10] to compute storage volumes.

Table 6-1. Storage properties of shapes

Shape	Area, A m² (ft²)	Wetted Perimeter m (ft)	Volume, V m³ (ft³)
Rect. (vert.)	BW (plan area)	n/a	Ad
Circ. (vert.)	$\dfrac{\pi D^2}{4}$ (plan area)	n/a	Ad
Trapez. (horiz.)	$d\left(\dfrac{d}{s_s}+W\right)$	$W+2\sqrt{d^2+\left(\dfrac{d}{s_s}\right)^2}$	AL
Segment of Circle	$\dfrac{D^2}{8}\left[2\cos^{-1}\left(1-\dfrac{2d}{D}\right)-\sin\left(2\cos^{-1}\left(1-\dfrac{2d}{D}\right)\right)\right]$	$D\cos^{-1}(1-\dfrac{2d}{D})$	AL
Rect. Wedge	dW	n/a	$0.5AL$
Trapez. Wedge	$d\left(\dfrac{d}{s_s}+W\right)$	n/a	$\dfrac{WsL^2}{2}+\dfrac{s^2L^3}{3s_s}$
Circ. Wedge (Ungula)	End area, A, per segment of circle	n/a	$L\dfrac{\frac{2}{3}a^3+cA}{D/2+c}$ where: $c=d-D/2$ $a=\sqrt{\dfrac{D^2}{4}-c^2}$

6.8.1 Sloping Conduit

The volume contained in a sloping conduit below a given water surface is not represented by the volume of a prism, but a wedge. Table 6-1 presents equations for wedges of rectangular and trapezoidal sections. In a circular cross section, the shape is referred to as ungula.

6.8.1.1 Ungula Formula

Table 6-1 includes an equation for the ungula volume. Variables are identified in Figure 6-5. The end area, A, is computed using the equation provided in Table 6-1 for a segment of a circle.

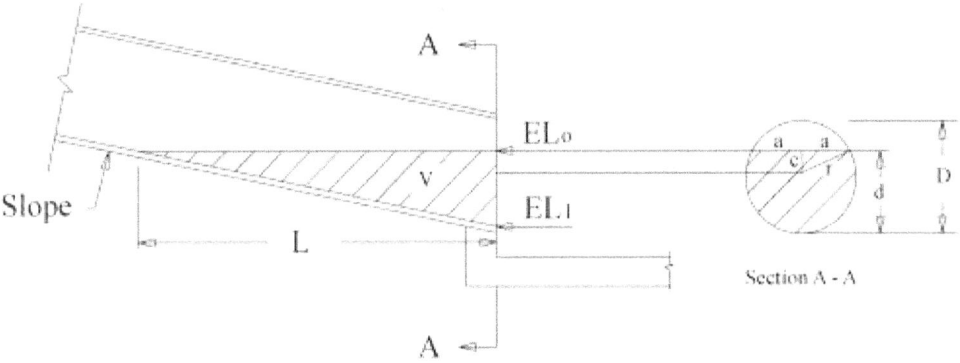

Figure 6-5. Ungula dimensions

6.8.2 Example of Storage in a Sloping Pipe

A 160 m (525 ft) long reinforced concrete pipe with a diameter of 1.2 m (3.9 ft) is placed at a slope of 0.004 m/m (0.004 ft/ft) and a downstream flow line elevation of 20.0 m (65.62 ft). Calculate the volume of water stored in the pipe at a depth of 0.5 m (1.64 ft).

Referring to Figure 6-5 and Table 6-1, the cross-sectional area at the downstream end of pipe is as follows.

Using the equation for the segment of a circle,

$$A = \frac{D^2}{8}\left[2\cos^{-1}\left(1-\frac{2d}{D}\right) - \sin\left(2\cos^{-1}\left(1-\frac{2d}{D}\right)\right)\right]$$

$$A = \frac{1.2^2}{8}\left[2\times\cos^{-1}\left(1-\frac{2\times 0.5}{1.2}\right) - \sin\left(2\times\cos^{-1}\left(1-\frac{2\times 0.5}{1.2}\right)\right)\right] = 0.446 \text{ m}^2 \text{ (4.8 ft}^2\text{)}$$

Referring to Section A-A in Figure 6-5,

$$c = d - D/2$$

$$c = 0.5 - 1.2/2 = (-)0.1 \text{ m } (-0.33 \text{ ft})$$

$$a = \sqrt{\frac{D^2}{4} - c^2}$$

$$a = \sqrt{\frac{1.2^2}{4} - (-0.1)^2} = 0.592 \text{ m } (1.94 \text{ ft})$$

Referring to Figure 6-5, the horizontal length of pipe inundated by water, L, is calculated as follows:

L = depth of flow/pipe slope

$L = 0.5/0.004 = 125$ m (410 ft)

Using the equation for the volume of an ungula,

$$V = L \frac{\frac{2}{3} a^3 + cA}{D/2 + c}$$

$$V = 125 \times \frac{\frac{2}{3} \times 0.592^3 + (-0.1) \times 0.446}{1.2/2 + (-0.1)} = 23.36 \text{ m}^3 \text{ (825 ft}^3\text{)}$$

6.8.3 Procedure to Establish Storage Unit Dimensions

Since the total storage is the sum of all storage in the collection system, storage unit and wet well below the maximum highwater and above the lowest pump elevation, it is necessary to size the unit to ensure that the total storage meets or exceeds the storage volume required.

Step 1.   Establish a flow line profile from the collection system to the sump that will be used for the storage unit.

Step 2.   Assuming the lowest pump elevation will be set at the invert of the storage unit at the sump entrance, compute the volume in the collection system (V_{cs}) below the allowable highwater.

Example: the collection system storage is ignored as being insignificant.

Step 3.   Assume a standard sump shape and size. Compute the sump volume between the maximum highwater and lowest pumping elevation (V_w).

Example: using a circular wet well with an inside diameter of 6.4 m (21 ft) and a maximum allowable highwater of 22 m (72.18 ft) and lowest pumping elevation of 20 m (65.62 ft) the volume in the sump is:
$V_w = \pi \times 0.25 \times 6.4^2 \times (22 - 20) = 64.3 \text{ m}^3 \text{ (2271 ft}^3\text{)}$

Step 4.   Choose a standard storage unit shape and section dimensions. Check local preferences and availability.

Step 5.   Compute the cross-section area of the chosen conduit, A_S.

Example: using a 1200 mm pipe, the cross-sectional area is:
$A_s = \pi \times 0.25 \times 1.2^2 = 1.13 \text{ m}^2 \text{ (12.2 ft}^2\text{)}$

Step 6.   Compute the length of storage unit required, L_S as:

$$L_{s(req)} = \frac{V_{req} - V_{cs} - V_w}{A_s} \tag{6-1}$$

where:

A_S	=	Cross-section area of conduit, m^2 (ft^2)
V_{req}	=	Minimum storage volume required, m^3 (ft^3)
V_{cs}	=	Volume in collection system below allowable highwater, m^3 (ft^3)
V_w	=	Volume in wet well below allowable highwater, m^3 (ft^3)
$L_{s(req)}$	=	Total length of storage unit required, m (ft)

For example, using the calculations above and an estimated required total volume of 260 m^3 (9182 ft^3), the estimated required length is:

$L_s = (260 – 0 – 64.3) / 1.13 = 173$ m (568 ft)

Step 7.   Determine the number of barrels of conduit required by dividing the total length required, $L_{S(req)}$, by the length available to fit the storage unit. For example, the longest storage unit length is 160 m:

number of barrels required = 173 / 160 = 1.08

Since the required storage is only estimated try only 1 barrel. Subsequent analysis will establish if the storage is sufficient.

6.8.4   Procedure to Develop Stage-Storage Curve

The stage-storage curve represents the total storage that is available within the system at any stage between the inlet elevation (invert) of the pump station and the maximum allowable elevation in the wet well, $HW_{s(max)}$. The total storage includes storage in the collection system, storage unit, and the wet well. The designer should compute storage values from the lowest pump stop elevation to the maximum allowable elevation. A level higher than the allowable is appropriate for evaluating the check storm.

The following steps indicate how to develop a stage-storage relationship.

Note: Some designers prefer to work in terms of actual elevation rather than stage.

Step 1.   Using the system profile plot, identify the lowest pumping elevation and set to zero stage. The available storage at this stage will be zero.

Step 2.   Choose an increment in stage.

Example: try 0.1 m (0.33 ft)

Step 3.   Establish a table of stage and storage for the full range of stages using the chosen increment.

Example: see Table 6-2.

Step 4.   For each stage, compute the storage as the sum of storage volumes in each element of the storage system at that stage. This is where the profile plot and layout are useful, especially for extensive storage networks. Record the computed storage for each stage in the aforementioned table.

Example: at a stage of 0.5 m, the volume in the wet well is:
$V_w = \pi \times 0.25 \times 6.4^2 \times 0.5 = 16.08$ m^3 (568 ft^3)

The volume in the collection system is assumed to be negligible

The volume in the storage unit is (see Section 6.8.2 - Example of Storage in a Sloping Pipe for calculation):
$V_w = 23.36$ m^3 (826 ft^3)

The total volume at 0.5 m stage is:
$V_t = 16.08 + 23.36 = 39.44$ m^3 (1393 ft^3)

Step 5.   Plot a curve of stage versus storage. The plot is especially useful in helping identify errors in the volume computations. The storage should increase with stage until $HW_{s(max)}$ is reached.

Figure 6-6 shows a completed stage versus storage curve using the value from Table 6-2.

Table 6-2. Example stage storage table

Stage (m)	Storage (m^3)		
	Sump	Storage Unit	Total
0	0.000	0.000	0.000
0.1	3.217	0.454	3.671
0.2	6.434	2.517	8.951
0.3	9.651	6.800	16.451
0.4	12.868	13.669	26.537
0.5	16.085	23.359	39.444
0.6	19.302	36.000	55.302
0.7	22.519	51.506	74.025
0.8	25.736	68.766	94.502
0.9	28.953	86.829	115.782
1	32.170	105.021	137.191
1.1	35.387	122.691	158.078
1.2	38.604	139.061	177.665
1.3	41.821	152.898	194.719
1.4	45.038	163.758	208.796
1.5	48.255	171.734	219.989
1.6	51.472	177.016	228.488
1.7	54.689	180.909	235.598
1.8	57.906	180.909	238.815
1.9	61.123	180.956	242.079
2	64.340	180.956	245.296
2.1	67.557	180.956	248.513
2.2	70.774	180.956	251.730
2.3	73.991	180.956	254.947
2.4	77.208	180.956	258.164
2.5	80.425	180.956	261.381

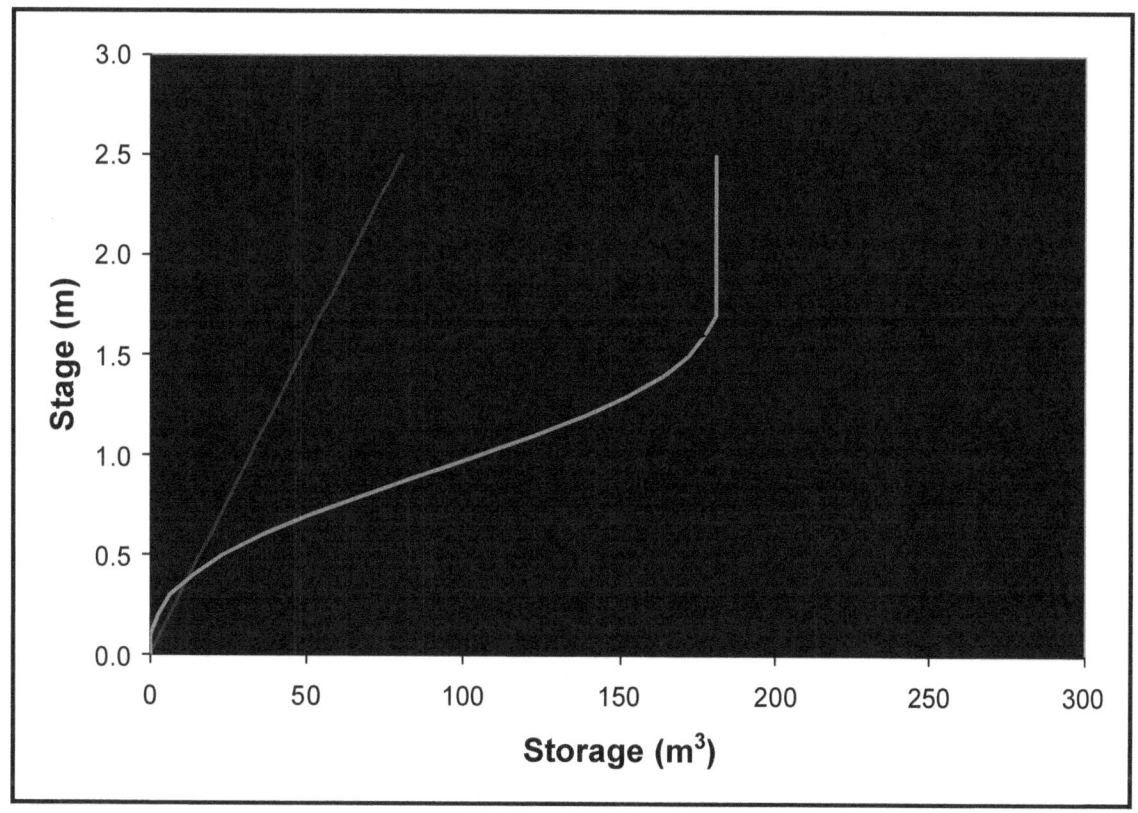

Figure 6-6: Example stage versus storage curve

This page intentionally left blank.

7. PUMP CONFIGURATION AND MASS CURVE ROUTING

7.1 INTRODUCTION

This chapter discusses considerations, criteria, and procedures for:

- pump configurations,
- discharge line components,
- total dynamic head,
- system curves, and
- mass curve routing.

7.2 CRITERIA AND CONSIDERATIONS FOR PUMP CONFIGURATIONS

7.2.1 Number of Pumps

The following table provides some quick-reference criteria for number of pumps.

Table 7-1. Suggested criteria for number of pumps

Criterion	Number of Pumps
Required Minimum	2
Preferred Minimum	3
Maximum	Not Specified

7.2.2 Size of Pumps

Generally, each main pump should have equal capacity and be of like type. Alternatively, it may be desirable to provide a low flow pump, sometimes termed nuisance pump, as the first to switch on. The remaining pumps would have a larger capacity. During low runoff events, only the small pump would be required, thus saving energy and reducing wear and tear. This is especially desirable when high total pumping capacity is provided by a few large pumps. The low flow pump may be turned off during peak flow events and is never included in the alternating start arrangement for the larger pumps.

7.2.3 Pump Sequencing

There are no specific criteria for pump sequencing. The number of sequencing alternatives is a function of the number of pumps; however, some general recommendations are appropriate.

- Use a low capacity pump as the first to switch on when using a small number of large capacity pumps. This pump is not part of the sequencing that follows.
- Provide automatically alternating sequences to distribute use of the larger pumps. This approach involves assigning a starting order for the pumps (say 1, 2, 3), then after one complete sequence the starting order is rearranged for the next operation (say 3, 1, 2), and so on. Otherwise, the first pump is always the most used and would require more frequent

Chapter 7 – Pump Configuration and Mass Curve Routing

maintenance and repair than the others. Also, this has the effect of increasing cycling time for a given storage volume or reducing the storage requirement for a given cycling time.

• The traditional sequence uses a cut-off order that is the reverse of the switch-on order. In this way, the first pump on is the last pump off.

The table below provides some sample starting sequences for systems with 2 to 4 pumps, the trend is clear for additional pumps.

No. of Pumps	First Sequence	Second	Third	Fourth	Fifth
2	1-2	2-1	1-2	2-1	1-2
3	1-2-3	3-1-2	2-3-1	1-2-3	3-1-2
4	1-2-3-4	4-1-2-3	3-4-1-2	2-3-4-1	1-2-3-4

The above sequencing is likely to provide the longest cycling times.

7.2.4 Pump Switching Elevations

The pump switching elevations should be set to ensure that the minimum cycling time will be greater than that specified by the manufacturer for a pump type and size. Refer to Section 7.2.9 - Pump Cycling Time for volume required to achieve minimum cycling time.

It is necessary to first establish the stop elevations for each pump in the station. Generally, unless specific configuration details dictate otherwise, the lowest pump stop elevation should be set at the invert elevation of the storage unit at the entrance to the sump. (If it is set higher, some of the storage unit will not be used during pumping.) Subsequent stop elevations can either be set at the same elevation as the lowest pump or more preferably staggered at predetermined heights above the previous pump stop level. Sometimes the height difference may be based on the sensitivity of the sensors used for switching (See Section 11.6.3 - Water Level Sensors).

Trial start elevations are established by determining the elevations at which the volume difference between the start and stop elevations for the particular pump is equal to the minimum cycle volume computed using Equation 7-6.

7.2.5 Procedure to Estimate Capacity and Number of Pumps

Use the following decision table to determine how to establish a configuration.

If the design is based on:	Then use alternate:	After which:
Fixed peak outflow, $Q_{P(req)}$	A	Proceed to usable storage and mass curve routing.
Fixed storage	B	Proceed to usable storage and mass curve routing.
Optimal pump capacity/storage	B	Proceed to usable storage and mass curve routing, but multiple iterations will be necessary.

7.2.5.1 Alternate A

For the trial storage system determined in Chapter 6. - Storage System, perform the following:

Step 1.   Compute the required total pumping in consistent units – m³/hour or gpm.

Step 2.   Select a trial number of pumps. Refer to Table 7-1.

Step 3.   Calculate the required capacity of each pump. The recommended design is to divide the total pumping rate by the number of pumps to establish equally sized pumps. For large pumps, a low flow pump may be necessary, but the main pumps should still be equally sized to handle 100% of the design flow.

7.2.5.2 Alternate B

For the trial storage system determined in Chapter 6. - Storage System, perform the following:

Step 1.   Plot the inflow hydrograph.

Step 2.   Referring to Figure 7-1, use the tangent on the rising limb of the inflow hydrograph to draw a line to a randomly selected point on the receding limb.

Step 3.   Estimate the volume contained between the line established in Step 2 and the top portion of the inflow hydrograph.

Step 4.   If the estimated volume is approximately equal to the total available storage provided, read the discharge rate at which the line intersects the receding limb. This is the target pump rate for the trial (or established) storage configuration. If the estimated volume is not approximately equal to the total available storage provided, return to Step 2.

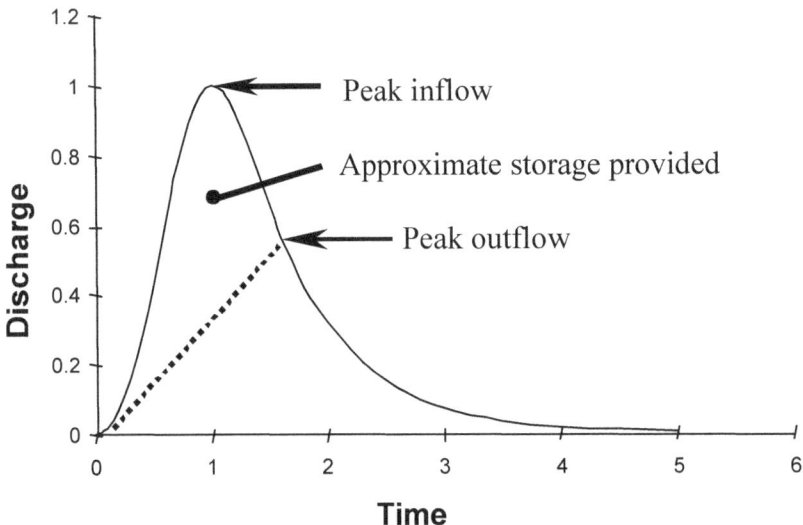

Figure 7-1. Estimating total pumping rate

7.2.6   Pump Cycling

Cycling refers to the time between starts of an individual pump. The shorter the cycling time, the more frequent a pump must start and stop. Each start of the pump requires a power surge. The power requirements increase with pump size. Also, the friction and vibration associated with starting the pump induces heat and wear. Repeated starting and stopping can result in overheating and excessive wear on the pump components. Therefore, it is necessary to keep cycle times as long as practicable.

Cycling is a function of storage and the pump discharge rate. For a given volume of storage, the cycling time reduces with increasing pumping rate. For a given pumping rate, the cycling time increases with increasing storage. For stormwater pumps, actual cycling times will vary in response to a particular runoff event. The characteristics of a runoff hydrograph will vary from one event to the next.

7.2.7   Allowable Minimum Cycling Time

The allowable minimum cycling time is dependent on the specific characteristics of the pump. Manufacturers provide specifications for the maximum number of starts per hour which translates to a minimum cycling time for individual pumps. For example, 10 starts per hour, or 6-minute cycles, may be quoted for a submersible pump. Generally, minimum cycling time increases with increasing pump capacity. Also, minimum cycling times vary with pump type: typically, minimum times for vertical motor-driven pumps are higher than for submersible pumps of similar capacity.

7.2.8   Usable Storage

There is a distinction between the volume of storage that is available for detaining the inflow hydrograph and the volume that is usable for pump cycling. See Section 6.3.3 - Available Storage. A lesser volume than the available storage is usable for cycling times because at any given moment as a pump turns on, some of the volume in the storage system is already occupied by water that is being conveyed to the pump. Under the assumption that the inflow rate has not changed when the pump turns off then on again, the same volume is still being used to convey water to the pump. Only the volume above this conveyance level is usable during pump cycling. Usable volume is a function of:

- sump dimensions,
- pump start elevation,
- pump stop elevation,
- storage system dimensions, and
- depth of flow in the storage system.

All unobstructed volume in the sump between a pump's start and stop elevations contributes to the usable volume. The volume increases as the difference between start and stop (pump range) increases. As the width of the storage system increases, the depth used to convey flow to the pump decreases and the usable volume increases.

7.2.9   Pump Cycling Time

7.2.9.1   First Pump

The following provides proof that, for a given pump with a capacity, Q_p, the cycling time will be a minimum when the inflow, Q_i, is equal to half of the pump capacity.

Assumptions:

- When pumping, the pump rate is constant and at pump capacity, Q_p
- The worst condition considered is a constant inflow rate, Q_i
- No pumping during filling cycle

The time to empty storage volume, V, is:

$$t = \frac{V}{Q_p - Q_i} \tag{7-1}$$

The time to fill the storage system (no pumping) is:

$$t = \frac{V}{Q_i} \tag{7-2}$$

If Q_i is expressed as a multiple of Q_p, $Q_i = xQ_p$, then the total cycle time is:

$$t = \frac{V}{Q_p - xQ_p} + \frac{V}{xQ_p} \tag{7-3}$$

For the minimum value of t,

$$\frac{dt}{dx} = -\frac{VQ_p}{(Q_p - xQ_p)^2} - \frac{V}{x^2 Q_p} = 0$$

Rearranging and dividing by V gives:

$$Q_p^2 x^2 + (Q_p - Q_p x)^2 = 0$$

For which,

$$x = 0.5$$

Thus,

$$Q_i = 0.5 Q_p$$

So, for the minimum cycle time, t, substituting for Q_i in Equation 7-1 gives:

$$t = \frac{4V}{Q_p} \tag{7-4}$$

If time is in minutes and flow rate is in cubic meters per second (cfs), the minimum cycle time is then described by Equation 7-5.

$$t = \frac{V}{15 Q_p} \tag{7-5}$$

Equation 7-5 can be used as a gauge to check cycling time for the first pump. However, since the inflow rate to a stormwater pump station is not constant, it is possible that the time to fill the usable storage could be shorter if the rainfall intensity and subsequent inflow rate increases significantly after the pump has evacuated the storage pit. If this occurs, the cycle time would be shorter than that established by Equation 7-5. The hypothetical worst-case cycle time would occur if the inflow stops just after the pump kicks in, and then immediately after the pump has evacuated the storage pit the inflow begins again at an inordinate rate such that the time to refill the pit is negligible. The cycle time would then reduce to $t = V/60Q_p$. Of course, this is highly unlikely, but it demonstrates that Equation 7-5 is not a strict minimum for a stormwater pump.

For design, it is usual to determine the required minimum volume for cycling by rearranging Equation 7-5 and by setting t to the manufacturer's recommended minimum cycle time:

$$V_{min} = 15Q_p t \qquad (7\text{-}6)$$

where:

Q_p	=	individual pump rate, m^3/s (cfs)
t	=	minimum cycle time, mins
V_{min}	=	minimum required cycle volume, m^3 (ft^3)

7.2.9.2 Cycling for Subsequent Pumps

The complexity of cycling checks can increase with the number of pumps. It is often assumed that if cycling criteria for the first pump are met, the subsequent pumps will have at least the same or longer cycling times. However, it can be shown that Equation 7-6 is valid for any number of equal-sized pumps when each volume, V_n is set equal to V_{min}, and is applied between the start and stop elevations of each pump as indicated in Figure 7-2. The proof is similar to that for one pump and is summarized as follows.

For the nth pump,

$$nQ_p > Q_i > (n\text{-}1)Q_p$$

Where:

n = number of pumps in the station,

Q_p = pumping rate of each individual pump,

Q_i = a worst-case constant inflow rate for the particular pump cycle, and

V_n = cycle volume for n^{th} pump (volume between start and stop of nth pump)

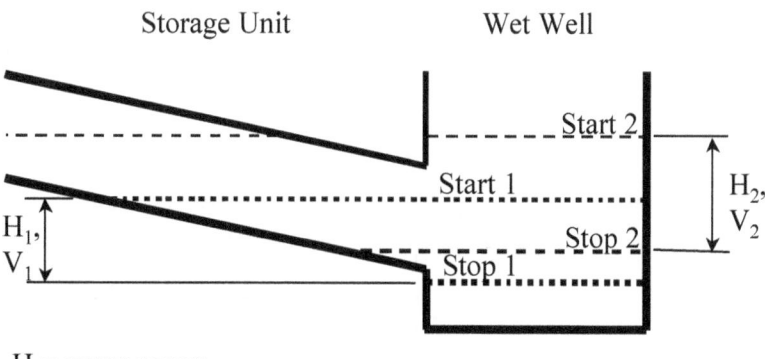

H = pump range
V = cycle volume
1, 2 = pump number

Figure 7-2. Cycling volumes

The cycle time for the nth pump is:

$$t = \frac{V}{nQ_p - xQ_p} + \frac{V}{xQ_p - (n-1)Q_p}$$

(7-7)

or:

$$t = \frac{V}{Q_p(n-x)} + \frac{V}{Q_p(x-n+1)}$$

Note that this is the time to fill and empty only the volume between the nth pump start and stop elevations, thus, once the nth pump stops, n-1 pumps are still operating.

For the minimum value of t,

$$\frac{dt}{dx} = \frac{V_n}{Q_p(n-x)^2} - \frac{V_n}{Q_p(x-n+1)^2} = 0$$

Rearranging and multiplying by Q_p/V gives:

$$(x-n+1)^2 - (n-x)^2 = 0$$

which reduces to:

$$2x - 2n + 1 = 0$$

or:

$$x = n - 0.5$$

Thus, the cycling time is a minimum for the second pump when the inflow rate is 1.5 times the individual pump rate, 2.5 times the individual pump rate for the third pump, and so on as summarized below.

Table 7-2. Critical inflow rates for pump cycling

Pump Number	Critical inflow rate
1	$0.5Q_p$
2	$1.5Q_p$
3	$2.5Q_p$
4	$3.5Q_p$

Substituting for x in Equation 7-7:

$$t = \frac{V}{Q_p(n-(n-0.5))} + \frac{V}{Q_p((n-0.5)-n+1)}$$

which reduces to:

$$t = \frac{4V_n}{Q_p} \text{ (t in seconds)}$$

or:

$$t = \frac{V_n}{15Q_p} \text{ (t in minutes)}$$

This rearranges to:

$$V_n = 15Q_p t$$

Thus, for any given number of pumps, the usable volume between the start and stop elevations for the pump in question must meet or exceed the value established in Equation 7-6.

7.2.10 Usable Storage Calculations

The usable storage for pump cycling is between the start and stop elevations of the pump under consideration less the volume below uniform depth (but above the stop elevation) in the collection system and storage system. The assumption here is that the inflow rate maintains uniform depth, d_u, in the collection line and storage unit after the pump shuts off, thus the volume between uniform depth and the flowline of the collection system and storage unit is unavailable. (See Figure 7-3). This is often assumed no matter what the configuration of the storage unit. The flow used for uniform depth is the critical flow for the particular pump as identified in Table 7-2.

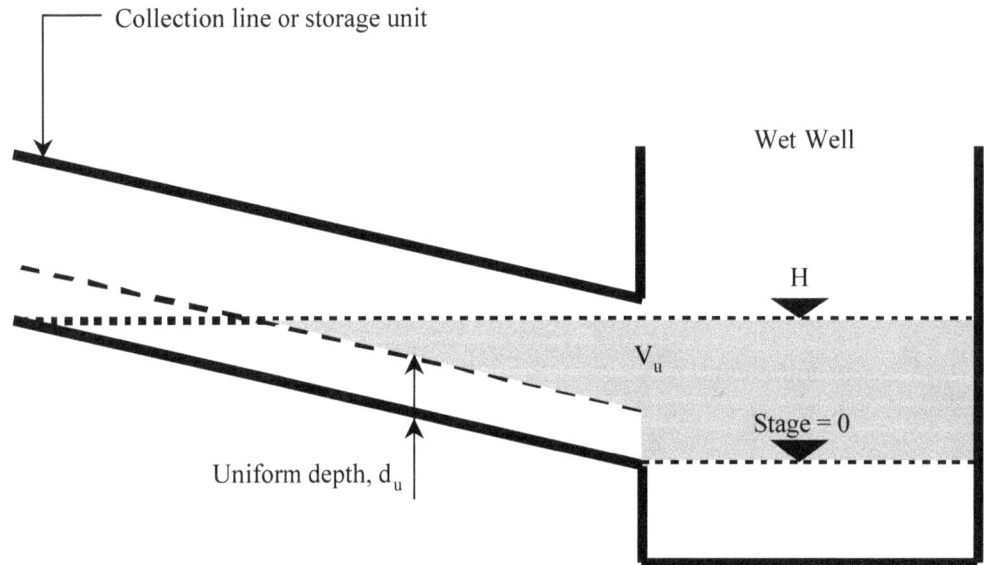

Figure 7-3. Usable storage at any stage

The volume equations shown in Section 6.8 - Storage Calculations apply when determining usable storage. However, it is essential that the appropriate portions of the wet well, storage unit and collection system are used. For any given reach of conduit, the usable portion is the volume in the conduit that is enclosed by the water level and the uniform depth line. The uniform depth in a conduit may be determined as follows:

Step 1.   Determine the critical discharge, Q, for the number of pump in question by referring to Table 7-2.
Step 2.   Choose a trial depth $d_{(trial)}$.
Step 3.   Compute the discharge, Q_u, for the trial depth using a form of Manning's Equation:

$$Q_u = \frac{C}{n}AR^{2/3}S^{1/2} \qquad\qquad (7\text{-}8)$$

where:

Q_u =   discharge, m^3/s (cfs)
C =   unit coefficient = 1 for SI (1.486 for English)
n =   Manning's roughness coefficient
A =   area of flow at uniform depth, m^2 (ft^2)
R =   hydraulic radius = area/wetted perimeter, m (ft)
S =   slope of conduit, m/m (ft/ft)
    a. Refer to HY-22[16] for a computer program to compute uniform depth, or
    b. Use Table 6-1 for the appropriate conduit shape to find equations for area and wetted perimeter.

Step 4.   Use the following table to decide how to proceed:

If:	Then:
$Q_u > Q$	Decrease $d_{(trial)}$ by using a value halfway between the current and previous values tried and return to Step 3
$Q_u < Q$	Increase $d_{(trial)}$ by using a value halfway between the current and previous values tried and return to Step 3
$Q_u = Q$	Proceed to Step 5

Step 5.   The current $d_{(trial)}$ represents uniform depth, d_u.

The volume under the uniform depth line may be computed as a prism of length, L. The length is the lesser of the conduit length and the distance between the downstream end of the conduit and the point at which the water level intersects with uniform depth (Figure 7-3). Refer to Table 6-1 for volumes of rectangular and trapezoidal sections and circular segments.

7.2.10.1 Example of Uniform Depth Calculation

A 160 m (525 ft) long storage unit for a two-pump facility is to comprise a 1.2 m (3.9 ft) reinforced concrete pipe at a slope of 0.004 m/m (0.004 ft/ft). The individual pump rate is 0.2 m³/s (7 cfs). Consider the first pump.

1.   The discharge at which the potential cycle time will be at a minimum is half of the pump rate for pump number 1: $Q = 0.2/2 = 0.1$ m³/s (3.5 cfs).

2.   Choosing a trial depth of 0.6 m (2 ft):

3. $Q_u = \dfrac{C}{n} A R^{2/3} S^{1/2} = \dfrac{1}{.013} \times 0.565 \times (\dfrac{0.565}{1.885})^{2/3} \times 0.004^{1/2} = 1.2$ m³/s (42.4 cfs)

4. $Q_u > Q$, therefore, reduce the assumed depth. After several iterations, repeating Step 3 for d = 0.165 m (0.54 ft) gives: $Q_u = \dfrac{1}{.013} \times 0.094 \times (\dfrac{0.094}{0.913})^{2/3} \times 0.004^{1/2} = 0.1$ m³/s (3.5 cfs)

5.   Since, $Q_u = Q$, uniform depth is 0.165 m (0.54 ft).

7.2.10.2 Stage versus Usable Storage

Using a similar approach to development of the stage-storage curve, the usable storage can be calculated by taking the available volume at the stage for the pump number under consideration and subtracting the volume in the collection system and storage unit that is below uniform depth for the critical flow rate (Table 7-2). Table 7-3 shows a usable storage table for the storage system used in Section 6.8.4 - Procedure to Develop Stage-Storage Curve. This approach is often precluded by ensuring that the cycling volume is provided in the sump between the start and stop levels for each pump.

Table 7-3. Example usable storage table

Stage (m)	Storage (m³)				
	Available			Total Usable	
	Sump	Storage Unit	Total	1 Pump	2 Pumps
0	0.000	0.000	0.000	0.000	0.000
0.1	3.217	0.454	3.671	3.217	3.217
0.2	6.434	2.517	8.951	6.434	6.434
0.3	9.651	6.800	16.451	9.651	9.651
0.4	12.868	13.669	26.537	12.868	12.868
0.5	16.085	23.359	39.444	16.085	16.085
0.6	19.302	36.000	55.302	19.302	19.302
0.7	22.519	51.506	74.025	62.006	47.632
0.8	25.736	68.766	94.502	81.814	64.370
0.9	28.953	86.829	115.782	100.733	83.459
1	32.170	105.021	137.191	122.141	104.648
1.1	35.387	122.691	158.078	143.028	125.535
1.2	38.604	139.061	177.665	162.615	145.122
1.3	41.821	152.898	194.719	179.670	162.176
1.4	45.038	163.758	208.796	193.747	176.253
1.5	48.255	171.734	219.989	204.940	187.446
1.6	51.472	177.016	228.488	213.439	195.945
1.7	54.689	180.909	234.601	219.551	202.058
1.8	57.906	180.909	238.815	223.766	206.272
1.9	61.123	180.956	242.079	227.029	209.536
2	64.340	180.956	245.296	230.246	212.753
2.1	67.557	180.956	248.513	233.463	215.970
2.2	70.774	180.956	251.730	236.680	219.187
2.3	73.991	180.956	254.947	239.897	222.404
2.4	77.208	180.956	258.164	243.114	225.621
2.5	80.425	180.956	261.381	246.331	228.838

7.2.11 Procedure to Determine Trial Switching

The switching scheme will affect the distribution of pumping with time. The implications of poor switching include excessive cycling and exceeding the allowable highwater level. The following approach establishes a trial scheme that is subsequently checked for both conditions.

Step 1. Establish the stop elevations (STOP) for all pumps. For the first pump, the stop elevation (Stop 1) should be set at the invert of the storage unit at wet well entrance or based on the typical wet well detail to be used. Suggested approaches for the subsequent stop elevations (Stop 2, Stop 3, etc) are either to stagger them at preset intervals of say 150 mm (0.5 ft) or 300 mm (1 ft), or to set all stops at the same level.

Example: Choosing 300 mm intervals, Stop 1 = 0 m (0 ft), Stop 2 = 0.3 m (1 ft)

Step 2. Determine the total volume at each stop elevation using the stage-storage curve.

Example: Referring to the example stage-storage curve established in Section 6.8.4 - Procedure to Develop Stage-Storage Curve,

> At Stop 1 = 0 m, the stop volume, S_{o1} = 0 m^3 (0 ft^3)
> At Stop 2 = 0.3 m, the stop volume, S_{o2} = 16.45 m^3 (581 ft^3)

Step 3. Estimate the minimum cycling time (t_c) for the main pump (largest) using pump manufacturer's information based on the trial pump size. For large submersible pumps a value of 6 minutes is typical. For vertical pumps, a value of 10 minutes is typical.

Example: Assume 6 starts per hour (to allow either a vertical pump or a submersible pump), which equates to a cycle time of 10 mins.

Step 4. Compute the minimum required usable storage using Equation 7-6.

Example: For a pump capacity of 0.2 m^3/s (7 cfs), the minimum cycling volume for each pump is:

> $V_{u(min)}$ = 15 x 0.2 x 10 = 30 m^3 (1059 ft^3)

Step 5. Establish a trial pumping range, δh, for the first main pump as follows:
   a. Determine the elevation at which the minimum cycling volume (usable volume) is met or just exceeded using the stage-storage curve, or the Usage Storage Table, Start 1.

Example: Referring to Table 7-3, the stage at which the usable storage for 1 pump exceeds 30 m^3 is 0.7 m (2.3 ft). This is Start 1.

   b. Calculate the trial range, δh, as the difference between Start 1 and Stop 1
Example: The trial range is 0.7 – 0 = 0.7 m (2.3 ft)
   c. If δh is less than 150 mm (0.5 ft), use 150 mm (0.5 ft), reset Start 1 to Start 1 + 150 mm (0.5 ft) and use the stage storage curve to determine the total storage at Start 1. A minimum difference of 150 mm (0.5 ft) is suggested to reduce the potential for concurrent activation of two or more pumps as a result of waves or other fluctuations in the water level.

Step 6. Establish remaining pump start elevations (Start 2, Start 3, etc.) as follows. For each remaining pump:
   a. Determine the usable volume at the stop elevation.

Example: Referring to Table 7-3, the usable volume at Stop 2 = 0.3 m (1ft) for two pumps operating is 9.65 m^3 (341 ft^3)

   b. Add the required cycling volume ($V_{u(min)}$) to the usable volume at the stop elevation to find the usable volume required for starting.

Example: Start available volume for Pump 2, S_2 = 30 + 9.65 = 39.65 m^3 (1400 ft^3)

c. Determine the elevation at which the required starting usable volume is met or just exceeded using the stage versus usable storage curve for the current number of pumps in operation. Set this as the preliminary start elevation. Figure 7-4 is a graphical representation of the relationship between usable volume and starting levels.

Example: Referring to Table 7-3, the stage at which a usable volume of 39.65 m^3 is exceeded is:

Start 2 = 0.7 m (2.3 ft)

d. If the start elevation is less than a minimum desirable height above the stop elevation, use the minimum desirable increment. This accounts for possible fluctuations due to turbulence and sensitivity of the sensors. A value in the range of 150 mm to 300 mm is typical.

Example: Since Start 1 is at 0.7 m, set Start 2 to 0.7 + 0.3 = 1.0 m (3.1 ft)

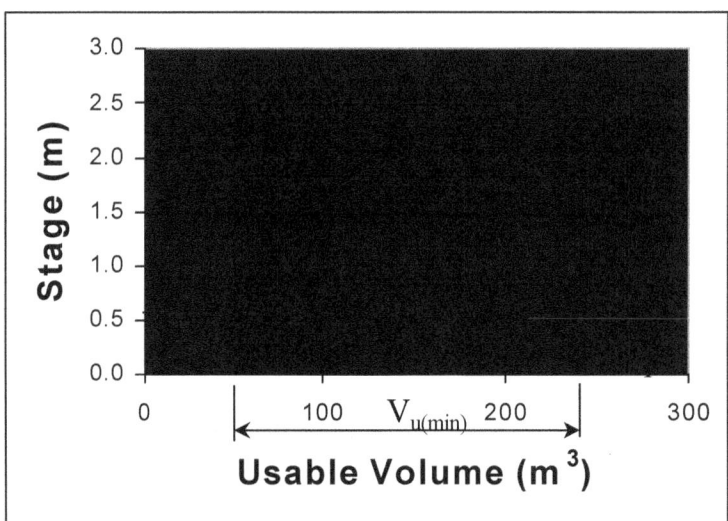

Figure 7-4. Usable volume and pump start levels

Step 7. Develop a table of switching levels (on and off) and available storage volumes. You can retrieve the available storage volumes from each elevation from the stage-storage curve derived in Section 6.8.4 - Procedure to Develop Stage-Storage Curve.

Example: The following table relates the trial switching levels and available volumes from Table 6-2.

Table 7-4. Pump switching table

	Pump 1		Pump 2	
On Stage - m (ft)	0.7	(2.30)	1.0	(3.28)
On Storage - m^3 (ft^3)	74	(2614)	137	(4845)
Off Stage - m (ft)	0.0	(0.00)	0.3	(0.98)
Off Storage - m^3 (ft^3)	0	(0)	16	(581)

Step 8. Develop a stage-discharge curve based on the pump switching table. Figure 7-5 shows a resulting plot for the switching table above and two pumps rated at 0.2 m^3/s (7 cfs).

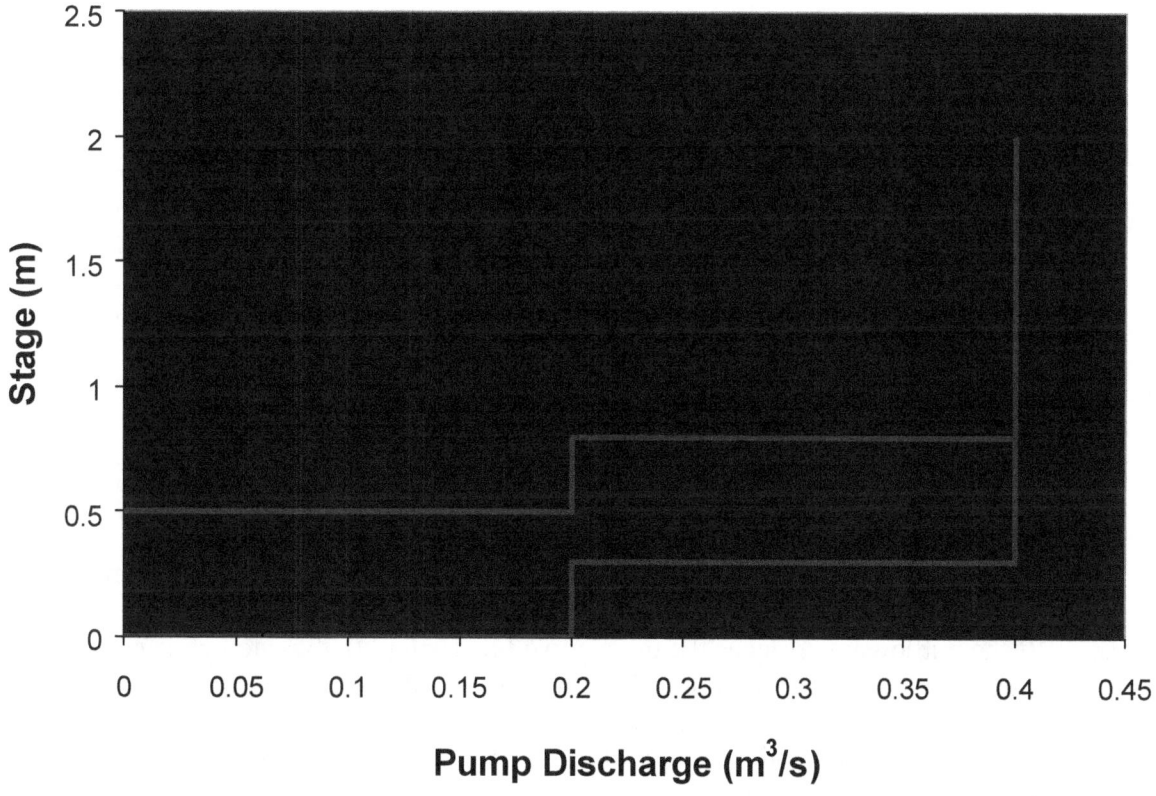

Figure 7-5. Sample stage-discharge curve

7.3 MASS CURVE ROUTING

The primary hydraulic evaluation of a trial pump station configuration is the determination of the cumulative outflow using the design mass inflow curve. The procedure to do this is called mass curve routing. The objectives of mass curve routing are to:

1. Determine the maximum available storage volume required by the trial pump size and switching and compare it with the trial available storage provided,

2. Determine the number of cycles a pump will operate.

No matter what design philosphy is employed, the mass curve routing cannot proceed until the following are established:

- inflow mass curve (refer to Section 5.5 - Procedure to Determine Mass Inflow),
- stage-storage curve (refer to Section 6.8.4 - Procedure to Develop Stage-Storage), and
- stage-discharge curve (refer to Section 7.2.11 - Procedure to Determine Trial Switching)

7.3.1 Procedure to Perform Mass Curve Routing

Refer to Figure 7-6 for the following steps which use the pumping operation established in Table 7-4 with two pumps rated at 0.2 m^3/s (7 cfs).

Step 1.   Plot the mass inflow curve (cumulative inflow versus time) on scaled paper. And set the storage to zero.

Step 2.   Identify the time at which the volume reaches the start volume for the first pump (point A on Figure 7-6). Draw a line at a slope that represents the pumping rate for one pump operating from this time at zero storage (a slope of 0.2 m^3/s for this example). This line begins the outflow curve.

Step 3.   If the pump rate line intersects the mass curve line, the pumping ceases. This is represented by drawing a horizontal line from the point of intersection. If the pump rate line does not intersect the mass curve line, jump to Step 5. For this example, you would need to jump to Step 5 because the mass inflow volume is increasing faster than the outflow.

Step 4.   Continue the horizontal line until the volume difference between the mass curve and outflow curve exceeds the start volume for the first pump again.

Step 5.   If the difference between the mass curve and outflow curve reaches the starting volume for the next pump (point B), draw a line that represents the discharge rate for an additional pump (0.4 m^3/s (14 cfs) for this example).

Step 6.   At any point, if the volume difference drops to one of the pump stop volumes, the rate line is reduced to reflect the lower pump rate (for example, point C).

Step 7.   If the outflow line reaches the mass curve line (point D), the rate line is reduced to zero (horizontal) and Steps 4 to 7 are repeated until the mass curve has ended.

Step 8.   Establish the maximum volume in storage by drawing a line that is tangent to the mass inflow curve and parallel to the maximum pumping rate during the curve routing. The vertical difference between the tangent point and the outflow curve represents the maximum required storage volume. In the example in Figure 7-6, the maximum storage volume is 223 m^3 (7875 ft^3). Note that, in the example shown in 6.3.4 Procedure to Estimate Total Available Storage Required, the provisional estimate of volume required was 260 m^3 (9182 ft^3).

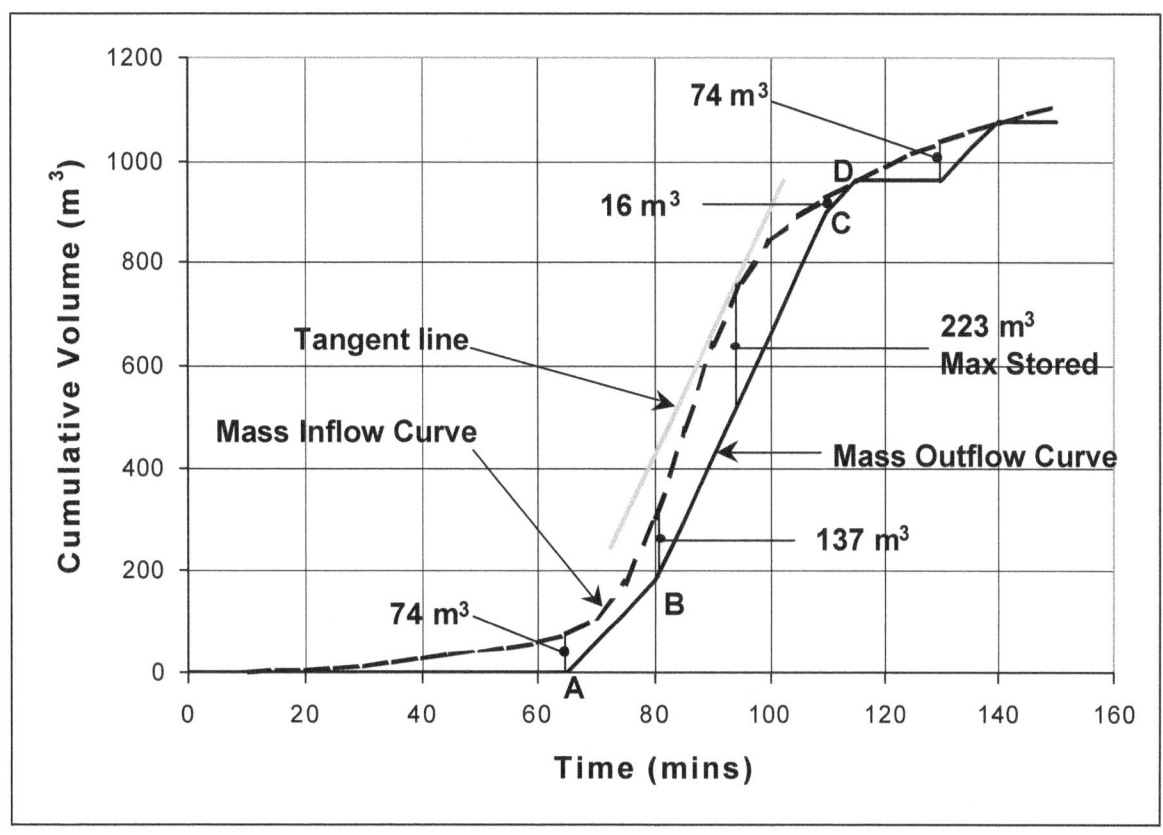

Figure 7-6. Example of graphical method for mass curve routing

7.3.2 Procedure to Check Routing Mass Curve

Step 1. Make note of the total storage volume provided, V_t, and the maximum required volume in storage, $V_{(req)}$, from the routing procedure.

Step 2. Review the mass curve routing to identify any periods in which more than one pump begins. If two pumps start within about 1 minute of each other, it will be preferable to adjust start elevations to increase the time between two pumps activating.

Step 3. Evaluate the number of times a pump must start and stop.

Step 4. Use the following table to decide how to proceed:

Condition	Likely cause	Suggested action
$V_{(req)} > V_t$ but cycle times long enough	<ul><li>Insufficient storage, or</li><li>Pumping capacity too low, or</li><li>Pumps not operating soon enough</li></ul>	<ul><li>Increase storage, or</li><li>Increase capacity or number of pumps, or</li><li>Reduce switch-on/ switch-off elevations</li></ul>
$V_{(req)}$ significantly lower than V_t but cycle times adequate	<ul><li>Pumping capacity too high</li></ul>	<ul><li>Decrease capacity or number of pumps</li></ul>
$V_{(req)} < V_t$ but cycle times too short	<ul><li>Insufficient storage between switching elevations</li></ul>	<ul><li>Adjust switching elevations to increase storage difference for most sensitive pump, or</li><li>Increase storage volume</li></ul>
$V_{(req)} < V_t$ and cycle times adequate	Pumping scenario is satisfactory	Proceed to next stage

Example: Referring to Figure 7-6 and Table 6-2:

- The maximum volume required is 223 m^3 (7875 ft^3).
- The total available volume (at maximum highwater stage of 2.0 m (3.3 ft) is 245 m^3 (8652 ft^3).
- The shortest time between starting of two pumps is from point A to point B - about 16 minutes.
- In a period of about 70 minutes (beginning at point A) Pump 1 cycles once for the design inflow.

8. DISCHARGE LINE AND PUMP SELECTION

8.1 DISCHARGE LINE COMPONENTS

For pump station design, the discharge line begins at the discharge end of the pump and ends at the beginning of the discharge conduit or outfall. A discharge line may include some combination of the following elements:

- Pipe
- Stop/check valves
- Flap gates/valves
- Elbows
- Manifolds
- Tee's
- Reducers
- Expanders
- Brackets, bolts and other fixtures

8.1.1 Basic Line

The discharge line should be kept as short and simple as possible. The simplest configuration is where each pump has its own discharge line, entirely independent of the other pumps. Each discharge line conveys pumped water from the pump to a channel or conduit outside the pump station. The elbow of the vertical riser from the pump should be set higher than the discharge line, with a slope down to the discharge end to minimize the volume of back flow when the pumps switch off. Where it is practicable, the centerline of each discharge pipe should be placed higher than the design backwater elevation in the receiving channel or conduit. A flap gate is generally preferred and should be placed at the terminus of each discharge line to prevent back flow if the centerline elevation at the end of the discharge pipe is below the design backwater elevation in the receiving structure. Consideration should also be given to the potential for back flow resulting from storms in excess of the design storm. A check valve may be desirable to prevent such back flow.

8.1.2 Manifold System

When excessive length and cost makes individual discharge lines impracticable, it is usual to connect the individual pump discharges into a common discharge line large enough to direct the combined discharge at an acceptable velocity. The connection element is called a manifold. Each pump discharge line must include a check valve to prevent recirculation of flow. It is rarely necessary to use a manifold system in highway pump stations.

8.1.3 Design Size

The size of the discharge pipe should be:

at least as large as the pump discharge diameter,
based on a suggested maximum discharge velocity of 3 m/s (10 fps), and
determined using the following equation:

$$D = 1.128\sqrt{\frac{Q}{V}} \qquad\qquad (8\text{-}1)$$

where:

D	=	pipe diameter, m (ft)
Q	=	discharge in pipe, m^3/s (cfs)
V	=	maximum velocity, m/s (fps)

8.1.4 Procedure to Determine Discharge Pipe Size

Step 1. Select a trial discharge line configuration. You will have an idea of pipe lengths required based on your pump location and proposed outfall. You will need to decide if you will use a manifold system or separate discharge lines and you will need to estimate the number of elbows, valves, and other appurtenances.

Example: Separate discharge lines are desired for two 0.2 m^3/s (7 cfs) pumps. Each line will need to be 20 m (66 ft) long with two 90 degree elbows.

Step 2. Select a maximum discharge velocity, V_d, say 3 m/s (10 fps), and compute the discharge pipe size using Equation 8-1:

Example: The target pipe diameter is:
D = 1.128 x (0.2 / 3)$^{0.5}$ = 0.29 m (0.95 ft)

Step 3. Round the diameter up to the nearest standard size.

Example: The nearest standard diameter for the target pipe size is 300 mm (12 in.)

8.1.5 Pump Discharge Pipe Material and Configuration Guidelines

Steel or ductile iron pipe are the primary choices for the pump discharge line. In the absence of other information, the following table provides guidance on pump discharge line configuration:

Table 8-1. Suggested pipe material selection

Line Length	Suggested Configuration	Recommended material
< 17m (50 ft)	Individual line per pump	Steel or ductile iron
17 m – 65 m (50 ft – 200 ft)	Individual line per pump	Steel or ductile iron inside station to concrete, corrugated metal, or plastic conduit outside station
> 65 m (200 ft)	Individual lines to common conduit or Manifold to single line	Steel or ductile iron inside station to concrete, corrugated metal, or plastic conduit conduit outside station Steel or ductile iron manifold in side station to concrete, corrugated metal, or plastic conduit outside station

8.1.5.1 Steel Pipe Grade

For typical stormwater pump stations, discharge heads are less than 20 m (50 ft). Associated pressures are relatively low so that a nominal 6 mm (1/4 inch) thickness of steel pipe wall will suffice. However, the stresses due to both internal pressure and external loading of backfill need to be checked. Pipe of ASTM A53 quality is satisfactory, or where fabrication is required, such as for a manifold, ASTM A36 steel is normally used. Although on occasions, ASTM A 283 Grade C has been specified for discharge lines and manifolds, it is not usually used if any premium price is involved compared with A36.

8.1.5.2 Steel Pipe Coating

Steel pipe and accessories require galvanizing or epoxy lining and coating as protection against corrosion, the former being preferable for interior and exterior coating of lines up to about sixteen-inch diameter, whether exposed inside the station structure or buried outside. Coal tar coating and pressure sensitive tape wrap are also used to protect buried steel pipe. For larger lines, the coatings mentioned are sometimes replaced by cement mortar, which is particularly applicable to manifolds buried outside the station.

8.1.5.3 Steel Pipe Joints

Buried flanged joints in steel pipe are frequently a source of leaks or internal corrosion in later years and should be avoided if possible by fabricating the discharge lines in one piece, or making field welded joints. Internal corrosion and rough surfaces increase the friction and head loss.

8.1.5.4 Concrete Pipe

For longer lines, concrete pressure pipe may be more economical than steel and is less subject to corrosion damage. Beveled pipe ends or elbow specials are standard with the concrete pipe industry, so that vertical or horizontal bends are readily laid. Thrust blocks to avoid separation of the line should be provided at bend points where resultant forces are significant. Joints should be of rubber-gasket type to withstand the internal pressure of the water and avoid leakage. The pipe wall thickness and pipe strength must be sufficient to withstand external pressures. Rubber-gasket jointed pipe is also very suitable where differential settlement of pump station and discharge line may be anticipated.

8.1.5.5 Other Materials

Ductile iron pipe can be considered for use as discharge lines if economically favorable. Outside of the station, options for conduit material selection include concrete, corrugated metal, and plastic.

8.1.6 Appurtenances

8.1.6.1 Couplings

The most common method of joining steel pipe is by forming or attaching a flange to each end of the pipe, and then by bolting the flanges together face-to-face, with a compressible gasket interposed between the flanges to form a water-tight joint.

For stormwater pump station construction, mild steel ring flanges of moderate rating are satisfactory. These have flat faces. AWWA (American Water Works Association) Standard C207-55, Class D, is an applicable specification, with such flanges being electrically-welded to the discharge piping. At least one pair of flanges is usually required in a wet-pit station, such as when a check valve is included in the discharge line. However, on occasions, flanges are not required.

Bolts and nuts (fasteners) required to make up flanges inside the pump station may be specified as carbon steel, but the exposed threads should be thoroughly cleaned and coated to prevent corrosion and remain workable in the event dismantling is required. The added cost of stainless steel fasteners is not usually warranted. Gasket material should be specified as cloth-impregnated neoprene.

8.1.6.2 Flexible Couplings

To facilitate assembly of the steel piping system and to allow for dimensional clearances and variations, it is necessary to include one or more flexible couplings in each line. These couplings are frequently known as Dresser couplings, relating to the name of the original manufacturer of couplings of this type. A typical covering appears in Figure 8-1.

In general, the coupling consists of two flange rings spaced apart by a sleeve and connected by a series of long parallel bolts around the periphery. The sleeve fits loosely over the adjoining spigot ends of the line pipe, and rubber ring gaskets are compressed between the flanges, the sleeve and the line pipe. The gaskets form a water-tight joint which is tolerant of some clearance and misalignment of the spigot ends; hence the description flexible. Just as the flexible coupling is essential for making up the system, it permits ready disassembly of the system when required. The flanges and sleeve should be epoxy coated (or galvanized). To prevent corrosion, the bolts (or fasteners) are often of stainless steel. The cost savings of mild steel or galvanized finish is a judgment factor for the individual designer.

The flexible coupling is very limited in its resistance to longitudinal forces or thrusts. It is often necessary to provide tie rods parallel to it on each side. These are sometimes more-or-less integral with the coupling and attached to the line pipe, but in a pump station it is often convenient to provide the same effect by tie-bolts between the pump and the structural concrete, again embracing the coupling. The sizing of the tie rods is determined by the cross-sectional area of the discharge line and the maximum discharge pressure of the pump. With proper tie rods or other restraint from the pump station structure, the flexible coupling will remain sealed and tight indefinitely while subjected to shock, vibration, pulsation or other adjustments of the discharge line.

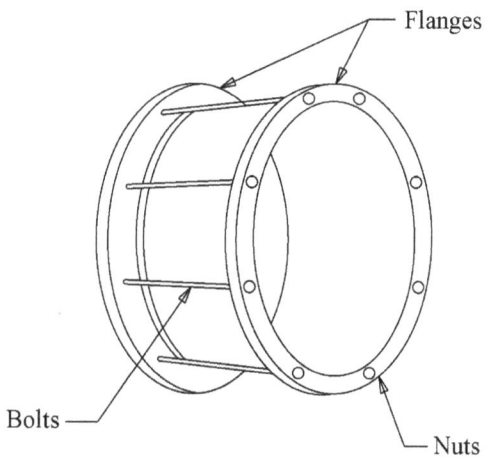

Figure 8-1. Typical flexible coupling

8.1.6.3 Steel-to-Concrete Adapter

A steel-to-concrete pipe adapter is required whenever such a change of material is included in the design. The adapter is often conveniently located in the concrete back wall of the pump station. This enables the steel adapter to be adequately anchored in the thickness of the wall, often with a circumferential ring or flange set on the center of the wall to serve also as a water-stop. Inside the station, the adapter may have either a steel flange or a spigot-end according to need. At some point in its length, the steel pipe size of the adapter is enlarged and at the downstream end a specially-sized bell ring is welded on. The sizing of the bell ring is performed by or for the concrete pipe manufacturer so as to match the spigot end of his concrete pipe with allowance for the necessary rubber gasket ring that ensures water-tightness under the discharge head. Internally, for a part of its length, the adapter is cement mortar lined. The remainder is epoxy coated.

The exterior of the adapter, which is inside the station, is also epoxy coated, while the portion which is embedded in the wall or projecting outside is coated with a cement mortar jacket.

8.1.7   Preventing Reverse Flow

Some reverse flow may be unavoidable and could be sufficient to cause reverse rotation of the pumps. The potential for this increases with discharge line size. It is essential to check with the pump manufacturer to determine if this is problematic and if so what mitigation measures may be appropriate, such as:

- a nonreverse ratchet fitted to the motor,
- a time-delay relay to prevent starting while the pump is running backwards, and
- an automatic check valve.

If engine-driven pumps are used, a non-reverse ratchet is required to prevent engine backfire.

8.1.8   Hydraulic Transients

A special problem may occur when a significant length of the discharge line from a manifold has an elevation above the pump discharge elbow. When pumps are shut down, the flow reversal toward the pumps could cause dangerous water hammer surges if regular check valves are used. A pump control valve is needed which opens and closes slowly and is controlled so that the pump always stops and starts against a closed valve. Under such a control system, the pump will operate at shutoff head for some seconds on start-up when water is in the line and the valve is opening. The shutoff power must be considered when sizing the drivers for such an installation. Pump curves usually show shut-off power requirements.

If vertical pumps are used and the maximum static head exceeds about 10 m (33 ft) an air vacuum discharge valve will be needed if the intake is not open to atmospheric pressure. In a stormwater pump, the intake is usually open to the atmosphere, in which case a release valve is not needed.

8.1.9   Accessories

Other accessories which may be required in the discharge system are air-release and vacuum valves, flexible couplings and tie rods, flange connections and fasteners, and steel-to-concrete pipe adapters.

8.1.10 Valves

A variety of valves are usually necessary and may include the following:

- air and vacuum valves,
- check valves, and
- flap gates.

The designer should always consult with the pump manufacturer to establish appropriate valve types and locations for the selected pumps. The following provides a general discussion and some guidance.

8.1.10.1 Air and Vacuum Valves

If a force main discharge line is used, or if high pressures in the discharge line are anticipated, it is necessary to provide an air release valve at the high point of a vertical pump discharge elbow. The pump column at the high point of the elbow will contain air that needs to be discharged to permit smooth starting of the pump and passage of a full body of water through the check valve. This valve should be the combination air and vacuum type, so that when the pump is shutdown, the valve also permits entry of air to avoid the harmful formation of a vacuum in the pump column. The sizing of these valves depends on the pump capacity and the volume of air to be released. Figure 8-2 provides an illustration of a ball float air vent.

On short discharge lines, if a flap gate is used at the outlet structure, a standard weight galvanized steel vent pipe installed vertically on top of the end of the discharge pipe at the outlet structure may be used in lieu of air-vacuum valves. The pipe should be capped on top to prevent entry of foreign objects, and should have holes drilled as airports in the upper part of the pipe before galvanizing.

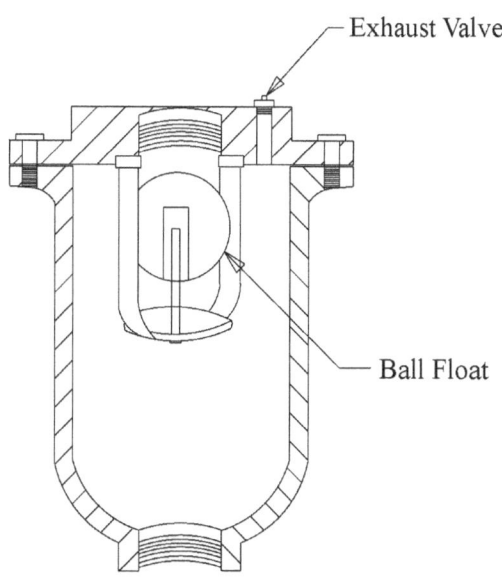

Figure 8-2. Ball float air valve

8.1.10.2 Check Valves

Check valves are used to prevent backflow into the pumps and subsequent recirculation. Any recirculation of flow can be harmful to the pump and wasteful of energy. There are two types of location at which a check valve is needed:

1. In the discharge lines adjacent to the main pumps, such as where the system includes a manifold.
2. In the discharge lines of sump pumps.

The check valves for the main pumps ensure that when any given pump is not operating, the flow of water being delivered to the manifold by other pumps cannot flow back through the non-operating pump. The flow velocity through the main check valves and consequently the head loss that they add to the system should be minimized. Therefore, it is usual to specify the main check valves to be the same size as the columns and discharge elbows of the vertical pumps they protect. The slanting-disc type with an oil dashpot feature, to ensure slow closing (non-slam) action of the flapper, is the most preferred type of check valve as shown in Figure 8-3.

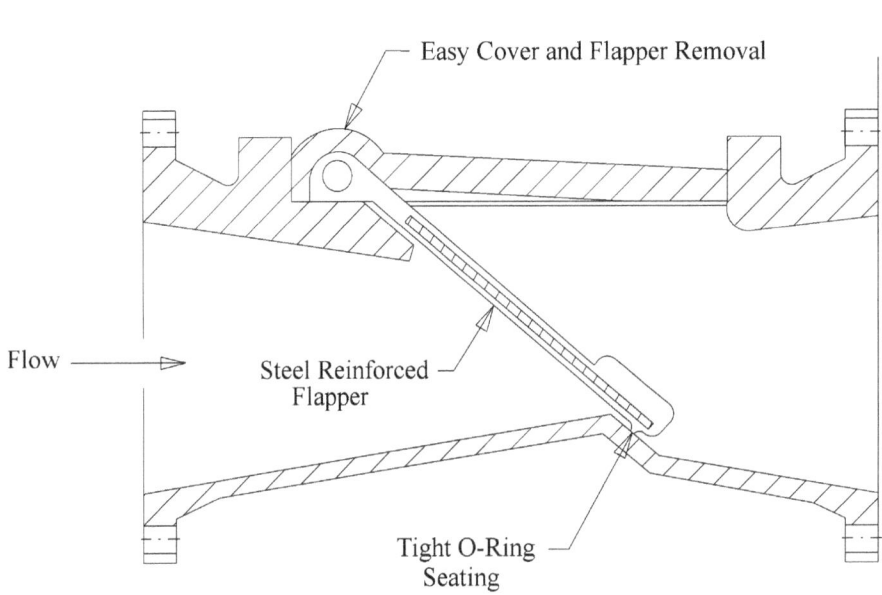

Figure 8-3. Rubber flapper swing check valve

The discharge lines of sump pumps are usually much smaller than the discharge lines of the main pumps. For sump pumps, any regular non-slam check valves with cast-iron body may be used, but the possibility of clogging is a consideration. A ball-type or rubber seated flapper-type check valve are preferred over the swing-disc type because they are less prone to jamming.

8.1.11 Flap Gates and Flap Valves

Flap gates and flap valves are intended to close the end of a pipe discharging into a receiving channel to prevent backwater from entering the discharge line.

8.1.11.1 Flap Gates

Typically, a flap gate consists of a heavy metal casting, or lid, circular in shape, hinged at the top from a body in such a manner as to close the end of the discharge pipe by gravity. See Figure 8-4. Water being pumped through the discharge line will raise the flap sufficiently to permit discharge. However, in so doing, there will be a head loss. Manufacturer's data for head losses should be used for design.

Potential problems with flap gates include:

- corrosion of the hinges resulting in poor operation,
- corrosion of gate and frame resulting in poor sealing ability, and
- debris can either prevent the gate from closing or prevent it from opening.

Where practicable, when using flap gates, the outlet should be above the flood level of the outfall. If the outfall is tidal and the pump outlet is lower than high tide elevation, conventional metal flap gates are not preferred.

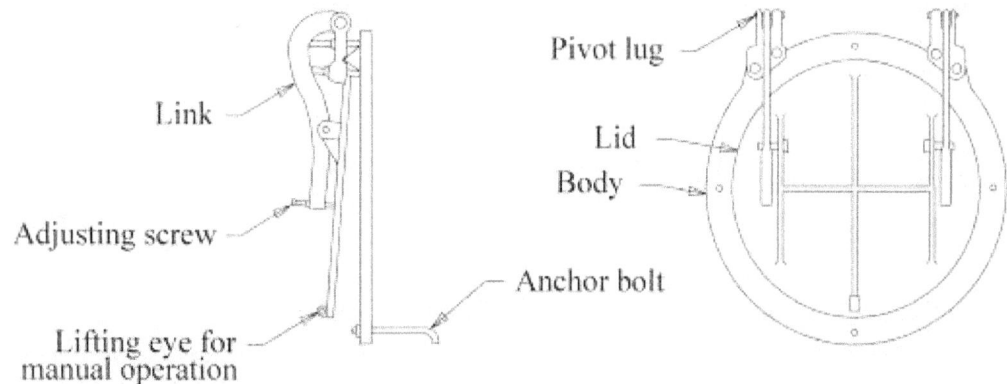

Figure 8-4. Typical flap gate

8.1.11.2 Flap Valves

Flap valves serve the same function as flap gates. A typical flap valve is shown in Figure 8-5. Commercially available flap valves are usually made out of neoprene. Generally, they require lower heads to operate than metal flap gates. They are more suitable in a submerged, saltwater environment than metal flap gates because of a lower propensity to corrode. Additionally, they are less prone to fail due to clogging than flap gates.

Figure 8-5. Rubber check valve

(Courtesy of Red Valve Company, Inc.)

8.1.12 Outfall

The discharge line terminates at the outfall. For most highway stormwater pump stations, the outfall should be one, or a combination of the following:

- open ditch or channel,
- stream or river,
- pond or lake,
- gravity storm drain conduit, and
- bay, estuary, or inlet.

Since design discharge line velocities will be high, about 3 m/s (10 fps), provision for permanent outlet protection should be considered. Figure 8-6 shows a typical outfall into an open channel. If the outfall is a storm drain conduit, it is likely that only a junction box will be necessary. Otherwise, the designer should refer to the FHWA publication *Design of Rip Rap Revetment, HEC-11*[14] for appropriate outfall protection measures such as concrete or flexible liners.

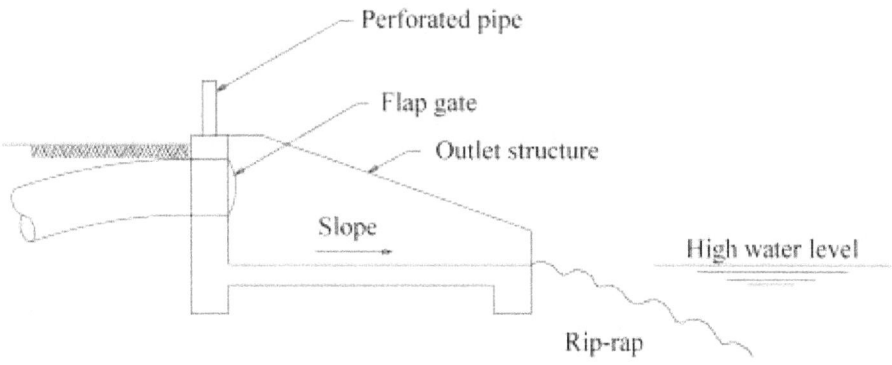

Figure 8-6. Typical outlet structure to channel

8.1.13 Friction Loss Equations

There still are differences in the way in which friction losses through pipes are calculated. These are:

- Darcy-Weisbach
- Hazen-Williams
- Manning's Equation

The choice is up to the designer and the available manufacturer's data. Generally, Hazen-Williams is used for the losses throughout the pump station and Manning's is used for the storm drain conduit outside the pump station.

8.1.13.1 Darcy-Weisbach

Probably the most accurate method is the Darcy-Weisbach method which uses the following equation:

$$h_f = \frac{fLV^2}{2gD} \qquad (8\text{-}2)$$

where:

h_f	=	friction head loss, m (ft)
f	=	friction factor
L	=	length of pipe, m (ft)
V	=	flow velocity, m/s (fps)
g	=	acceleration due to gravity, m/s^2 (ft/s^2)

The friction factor, f, is a function of Reynold's Number and flow type (turbulent or laminar). The factor is selected using a Moody Chart. The disadvantage of using this method is that the friction factor varies with pipe size, type and velocity of flow. As a result, this method is seldom used. However, if flow is in the turbulent range, friction factors can be considered constant for a given pipe diameter.

8.1.13.2 Hazen-Williams

The Hazen-Williams equation for friction loss, the most widely used equation, appears as follows:

$$h_f = \frac{C_u V^{1.85} L}{C^{1.85} D^{1.165}}$$ (8-3)

where:

h_f	=	friction loss, m (ft)
L	=	length of pipe, m (ft)
C_u	=	unit conversion coefficient = 6.83 SI (3.022 English)
V	=	discharge velocity, m/s (cfs)
C	=	Friction factor
D	=	pipe diameter, m (ft)

The Hazen-Williams equation should only be used for turbulent flow and is most applicable to water at a temperature of about 16 C (60 F). The friction factor, C, for the Hazen-Williams varies with pipe material and is typically in the range of 60 to 160. A design value of 100 is typical for smooth steel pipe and smooth concrete pipe.

8.1.13.3 Manning's Equation

Manning's Equation can be used in the following form to compute friction head loss:

$$h_f = C \frac{V^2 n^2}{R^{4/3}} L$$ (8-4)

where:

h_f	=	friction loss, m (ft)
C	=	unit conversion factor, 1 SI (0.453 English)
V	=	flow velocity, m/s (fps)
n	=	Manning's roughness coefficient
R	=	hydraulic radius = area/wetted perimeter, m (ft)
L	=	length of conduit, m (ft)

For a circular conduit flowing full, the above simplifies to:

$$h_f = 0.75 C \frac{V^2 n^2}{D^{4/3}} L$$ (8-5)

where:

D	=	pipe diameter, m (ft)

This method is considered less accurate than Darcy-Weisbach, but the roughness coefficient does not vary with velocity, making it more amenable. Typical design values of roughness (n) are 0.013 for concrete and 0.011 for smooth steel. Generally, Manning's Equation is applied to the storm drain conduit in the collection system and beyond the pump station discharge line.

8.1.14 Appurtenance Energy Losses

The most common approach to computing energy losses through appurtenances such as valves and elbows is by use of a dimensionless loss factor, K, applied to the velocity head as follows:

$$h_l = K\frac{V^2}{2g} \tag{8-6}$$

where:

h_l = friction loss through appurtenance, m (ft)
K = loss factor based on standard data or manufacturer's specified data
V = velocity through appurtenance, m/s (fps)
g = acceleration due to gravity, m/s^2 (ft/s^2)

Where an appurtenance incurs a velocity change, such as a reducer or expansion, the head loss calculation takes the following form:

$$h_l = K\frac{\left(V_2^2 - V_1^2\right)}{2g} = K\frac{\delta(V^2)}{2g} \tag{8-7}$$

where:

h_l = friction loss through appurtenance, m (ft)
K = loss factor based on standard data or manufacturer's specified data
V_1 = entrance velocity to appurtenance, m/s (fps)
V_2 = exit velocity from appurtenance, m/s (fps)
g = acceleration due to gravity, m/s^2 (ft/s^2)

Table 8-2 presents a means of estimating loss coefficients for typical appurtenances which may be used in the absence of manufacturer's specified losses. Note that dual units are given. The coefficients are based on a diameter given in millimeters for metric units and inches for English units. The exponent, b, is dimensionless.

Table 8-2. Loss coefficients for pipe appurtenances
Adapted from reference 17.

Element	Description	Coefficients			K Form	Loss Form
		Metric	English			
		a (D in mm)	a (D in inches)	b	$K =$	$h_f =$
	Regular flanged 90° Elbow	0.887	0.430	-0.224	aD^b	$KV^2/2g$
	Long radius flanged 90° Elbow	3.134	0.435	-0.610	aD^b	$KV^2/2g$
	Long radius flanged 45° Elbow	0.578	0.220	-0.298	aD^b	$KV^2/2g$
	Flanged Return Bend	1.080	0.450	-0.271	aD^b	$KV^2/2g$
	Flanged Tee - Line Flow	2.283	0.450	-0.502	aD^b	$KV^2/2g$
	Flanged Tee - Branch Flow	1.072	0.260	-0.438	aD^b	$KV^2/2g$
	Flanged Gate Valve	2.488	1.000	-0.282	aD^b	$KV^2/2g$
	Flanged Swing Check Valve	2.000	2.000	0.000	2	$KV^2/2g$
	Restrictor	n/a	n/a	n/a	0.1	$K \delta(V^2/2g)$
	Enlarger	n/a	n/a	n/a	0.2	$K \delta(V^2/2g)$
	Sudden Enlargement	n/a	n/a	n/a	0.5	$K \delta(V^2/2g)$

8.1.14.1 Example of Losses through Discharge Line

A discharge line consists of 20 m (65.6 ft) of 300 mm (12 in.) steel pipe with two long radius flanged 90 degree elbows and a flanged swing check valve. The pumping rate is 0.2 m³/s (7 cfs).

1. The discharge velocity is:

$$V = \frac{Q}{A} = \frac{0.2}{\frac{\pi}{4} \times 0.3^2} = 2.83 \text{ m/s (9.3 fps)}$$

2. Using Equation 8-3, the friction loss through the pipe is:

$$h_f = \frac{C_u V^{1.85} L}{C^{1.85} D^{1.165}} = \frac{6.83 \times 2.83^{1.85} \times 20}{100^{1.85} \times 0.3^{1.165}} = 0.759 \text{ m (2.49 ft)}$$

3. Referring to Table 8-2, the loss expression for a long radius flanged 90 degree elbow is:

$$h_l = aD^{-0.61} \frac{V^2}{2g} = 3.134 \times 300^{-0.61} \times \frac{2.83^2}{2 \times 9.81} = 0.039 \text{ m (0.13 ft)}$$

4. The loss expression for a flanged swing check valve is:

$$h_l = K \frac{V^2}{2g} = 2 \times \frac{2.83^2}{2 \times 9.81} = 0.816 \text{ m (2.68 ft)}$$

5. The total losses through the discharge line are found by summating the individual losses:
Total loss = $0.759 + 2 \times 0.039 + 0.816 = 1.653$ m (5.42 ft)

8.2 TOTAL DYNAMIC HEAD AND SYSTEM CURVE

8.2.1 Total Dynamic Head

Total dynamic head, *TDH*, represents the total energy required to raise the liquid from the intake to the discharge point. It comprises four components:

*   static head,
*   friction head,
*   velocity head, and
*   pressure head.

Figure 8-7 shows the components of total dynamic head and the following equation presents the relationship:

$$TDH = H_s + \Sigma h_f + h_v + \delta H_p \tag{8-8}$$

where:

$$
\begin{aligned}
H_s &= \quad \text{static head, m (ft)} \\
\Sigma h_f &= \quad \text{total head losses in pump and discharge line, m (ft)} \\
h_V &= \quad \text{velocity head, m (ft)} \\
\delta H_p &= \quad \text{pressure head change between outlet and intake, m (ft)}
\end{aligned}
$$

If the inflow and outflow are open to the atmosphere, the pressure head change term, δH_p, will be zero. This is usually the case for stormwater pump stations.

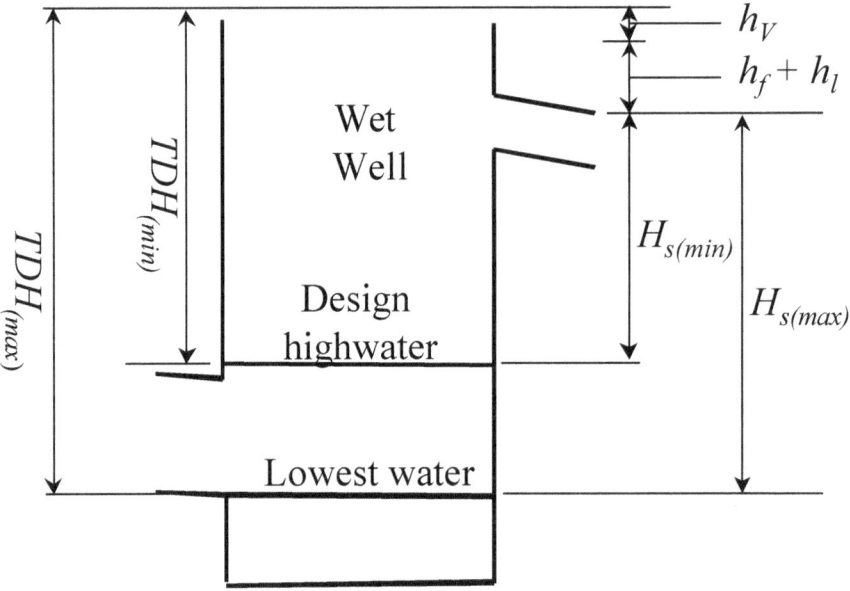

Figure 8-7. Components of total dynamic head

8.2.2 System Head Curve

A system head curve represents the variation in total dynamic head with respect to pumping rate. At zero flow, the total dynamic head is equal to the total static head. As the pumping rate increases, the velocity head, friction losses, and pump losses increase. Thus, the total dynamic head increases with pumping rate as indicated in Figure 8-8.

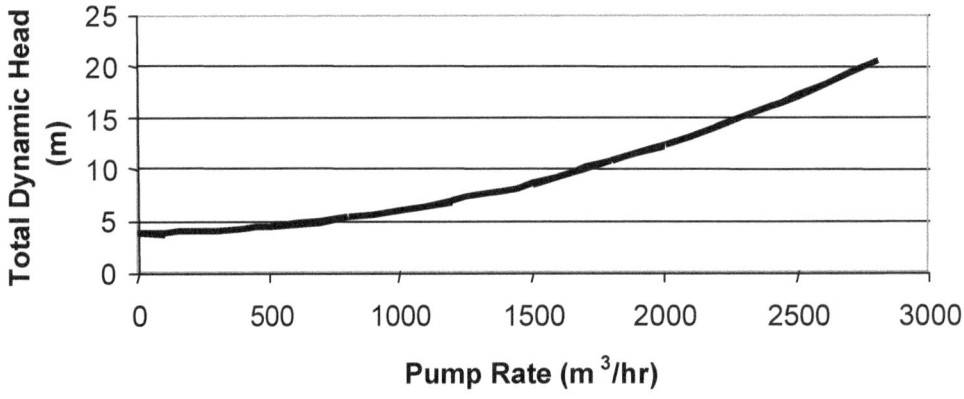

Figure 8-8. Typical system head curve

For highway stormwater pump stations, the static head will vary between the lowest elevation from which the stormwater is pumped and the maximum storage elevation. There will be minimum and maximum system curves as seen in Figure 8-9.

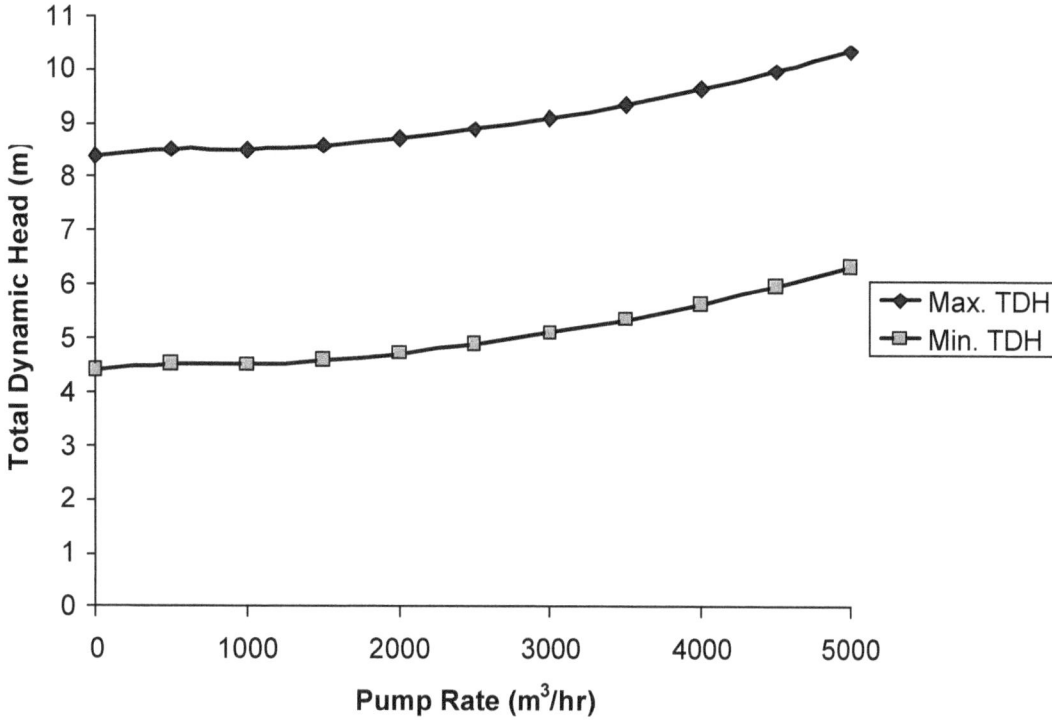

Figure 8-9. System curves for maximum and minimum static heads

8.2.3   Procedure to Determine System Curves

When selecting specific manufacturer's pumps and piping:

- the pump selection is dependent on the system head curve and power requirements,
- the power requirements are dependent on the total dynamic head requirements,
- the system head curve is dependent on total dynamic head,
- the total dynamic head is dependent on the pump and pipe head losses, and
- the head losses are dependent on the selected pumps and piping.

The designer must choose which way to proceed. The method presented here begins by estimating the system curves before selecting manufacturer's products. The assumptions are then checked for validity after selection.

Step 1.   Determine the maximum static head, $H_{s(max)}$. This is the difference in height between the outflow head level or discharge pipe elevation and the lowest pumping elevation (lowest pump stop elevation). Use the following table to determine the outflow head level.

If the centerline of the discharge pipe is:	then set the outflow head level to:	Comment
lower than the estimated backwater from the receiving water (outfall),	the estimated backwater level from the receiving water.	Not preferable, but if so, a flap gate will definitely be necessary.
Lower than normal depth of flow in the outfall,	normal depth in the outfall.	As above.
Higher than both normal depth of flow and backwater in the outfall,	centerline level of discharge pipe.	This is the preferred condition, where practicable.

Example: A 2-pump system has a lowest pumping elevation of 20 m (65.6 ft), a maximum highwater in the pump station of 22 m (72.7 ft), a centerline level of discharge pipe of 30 m (98.4 ft) and has a free outfall. The sump is at atmospheric pressure. The maximum static head is:
$$H_{s(max)} = 30 - 20 = 10 \text{ m (32.8 ft)}$$

Step 2. Determine the minimum static head, $H_{s(min)}$. This is the difference in height between the outflow head level and the maximum highwater level in the wet well. The same conditions described in Step 1 apply to the outflow head level.

Example: The minimum static head is:
$$H_{s(min)} = 30 - 22 = 8 \text{ m (26.3 ft)}$$

Step 3. Select a starting discharge, Q, that is greater than zero but lower than the target pump rate (Q_p).

Example: Assuming previous values already computed, use a discharge,
$$Q = 0.2 \text{ m}^3/\text{s (7 cfs)}.$$

Step 4. Compute the actual pipe velocity using the continuity equation as follows:

$$V = Q / A \tag{8-9}$$

where:
$$V \quad = \quad \text{pipe velocity, m/s (fps)}$$
$$Q \quad = \quad \text{discharge, m}^3/\text{s (cfs)}$$
$$A \quad = \quad \text{pipe sectional area, m}^2 \text{ (ft}^2\text{)}$$

Example: Using a 300 mm pipe, the pipe velocity is:
$$V = 0.2 / (0.25 \times \pi \times 0.3^2) = 2.83 \text{ m/s (9.3 fps)}$$

Step 5.   Compute the velocity head, h_v, using the following equation:

$$h_V = \frac{V^2}{2g} \qquad\qquad (8\text{-}10)$$

where:

h_V	=	velocity head, m (ft)
g	=	acceleration due to gravity, 9.81 m/s^2 (32.2 ft/s^2)

Example: The velocity head is:
$h_V = 2.83^2 / (2 \times 9.81) = 0.41$ m (1.34 ft)

Step 6.   Compute the head losses through pump discharge elements. Refer to Section 8.1.14.1- Example of Losses through Discharge Line.

Example: Referring to Section 8.1.14.1 - Example of Losses through Discharge Line, the sum of pipe losses and appurtenance losses in the discharge line at $Q = 0.2$ m^3/s (7 cfs) is 1.65 m (5.4 ft).

Step 7.   Compute TDH_{min} and TDH_{max} for the minimum and maximum static heads, $H_{s(min)}$ and $H_{s(max)}$, respectively, using Equation 8-8. Assuming the inlet and outlet are open to the atmosphere, δH_p will be zero

Example: The maximum total dynamic head is:
$$TDH_{(max)} = H_{s(max)} + \Sigma h_f + h_v = 10 + 1.65 + 0.41 = 12.06 \text{ m (39.6 ft)}$$
The minimum total dynamic head is:
$$TDH_{(min)} = H_{s(min)} + \Sigma h_f + h_v = 8 + 1.65 + 0.41 = 10.06 \text{ m (33.0 ft)}$$

Step 8.   Compute the arithmetic average (TDH_{ave}) of the minimum and maximum total dynamic heads.

Example: The average total dynamic head is:
$TDH_{(ave)} = (10.06 + 12.06) / 2 = 11.06$ m (36.3 ft)

Step 9.   Using the full range of Q, repeat Step 4 to Step 8 until you have developed the TDH information for the full range of flows.

Example: Table 8-3 shows the results of repeating the total dynamic head computations for a range of discharges.

Table 8-3. Example TDH calculations

Pump Rate	Min. Static Head	Min. Static Head	Velocity	Velocity Head	Pipe Friction Loss	Bend Losses	Transition Losses	Valve loss	Min. TDH	Max. TDH	Ave. TDH
m³/s	(m)	(m)	(m/s)	(m)	(m)	(m)	(m/s)	(m)	(m)	(m)	(m)
0	8	10	0.000	0.000	0.000	0.000		0.000	8.00	10.00	9.000
0.05	8	10	0.707	0.026	0.058	0.005		0.051	8.14	10.14	9.140
0.1	8	10	1.415	0.102	0.211	0.020		0.204	8.54	10.54	9.536
0.15	8	10	2.122	0.230	0.446	0.044		0.459	9.18	11.18	10.179
0.2	8	10	2.829	0.408	0.759	0.079		0.816	10.06	12.06	11.062
0.25	8	10	3.537	0.638	1.147	0.123		1.275	11.18	13.18	12.183
0.3	8	10	4.244	0.918	1.607	0.177		1.836	12.54	14.54	13.539

Step 10.   Plot the TDH_{max}, TDH_{min}, and TDH_{ave}, versus discharge. This is the system curve plot.

Example: Figure 8-10 shows the system curve plot for the values in Table 8-3.

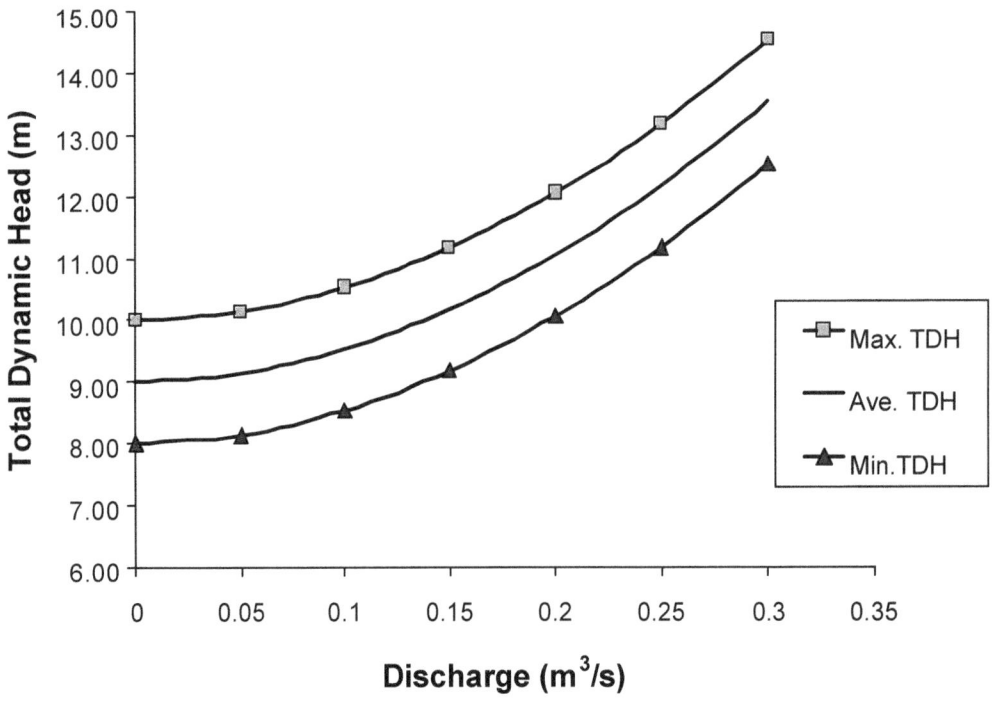

Figure 8-10. Example system curve

Step 11.   Establish the target design variables TDH_{ave}, the average of TDH_{max} and TDH_{min}, for the target pump capacity, Q_p. (That is, the capacity of one pump, not total pumping rate).

Example: Referring to Table 8-3, at a design pump rate of 0.2 m³/s (7cfs):
TDH(max) = 12.06 m (39.6 ft)
TDH(min) = 10.06 m (33.0 ft)
TDH(ave) = 11.06 m (36.3 ft)

8.3 PUMP CHARACTERISTICS

8.3.1 Cavitation

Cavitation is a hydraulic phenomenon in which vapor bubbles form and suddenly collapse (implode) as they move through a pump impeller. Implosions occur on each of the vanes of the impeller causing excessive noise. The hydraulic effect on the pump is a significant reduction in performance. The mechanical effects can include shock waves and vibration which may result in damage to the impeller vanes, bearings, and seals.

Cavitation occurs when the pressure in the liquid is reduced to the liquid's vapor pressure such that boiling begins to occur, even though the liquid's temperature may not have changed. Therefore, to prevent cavitation, it is necessary to ensure that the pressure does not drop to the liquid's vapor pressure.

The primary causes of cavitation in highway stormwater pump stations include the following:

- the impeller vane travels faster than the liquid,
- suction is restricted,
- the required NPSH is either equal to or greater than the available NPSH, and
- the specific speed is too high for optimum design parameters.

8.3.2 Vortexing

A vortex can occur at the impeller and may extend to the surface of the liquid. If this occurs, air will be sucked into the pump. The effects can be similar to cavitation: reduced hydraulic efficiency and increased wear on the pump. Dicmas identifies three stages of vortexing[3]:

- Type I – first inception of vortexing in which bubbles of air are drawn towards the pump
- Type II – vortexes form for a period of less than 30 seconds, pulling in air and possibly debris, and
- Type III – continuous vortexing draws in large volumes of air and possibly debris.

Figure 8-11 shows Type I and:Type III vortexing. Type II is a combination of Type I and Type III.

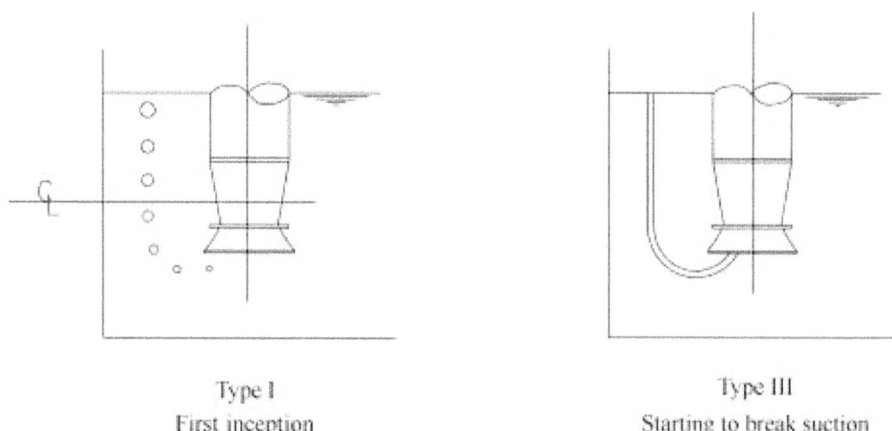

| Type I | Type III |
| First inception | Starting to break suction |

Figure 8-11. Types of vortex

It is preferable to avoid vortexing. However, Type I vortexing will generally not harm the pump and no significant reduction in performance is expected and thus can be deemed acceptable. Type II vortexing can reduce pump performance intermittently and Type III vortexing can cause serious reductions in pump performance along with cavitation, noise and vibration. The potential for vortexing is limited by providing adequate submergence (See 8.3.6 - Submergence).

8.3.3 Net Positive Suction Head Required

Net positive suction head required (NPSHR) is the head above vapor pressure head required to ensure that cavitation does not occur at the impeller. The value is specific to each pump inlet design. It is independent of the suction piping system. The NPSHR is determined by bench scale tests of geometrically similar pumps operating at a constant speed and discharge with varying suction heads. The development of cavitation is usually indicated by approximately a 3% drop in the head developed at the suction inlet valve. The NPSHR is specified by the manufacturer.

8.3.4 Net Positive Suction Head Available

Net positive suction head available (NPSHA) is the head available above vapor pressure head to move a liquid into the impeller unit of the pump. It is necessary to ensure that the NPSHA exceeds the NPSHR to prevent cavitation. The following equation is used to compute NPHSA:

$$NPSHA = H_{pa} + H_s - h_f - H_{vp} \tag{8-11}$$

where:

H_{pa} = the atmospheric pressure head on the surface of the liquid in the sump – m (ft).

H_s = static suction head of liquid. This is height of the surface of the liquid above the centerline of the pump impeller – m (ft)

h_f = total friction losses in the suction line – m (ft)

H_{vp} = the vapor pressure head of the liquid at the operating temperature – m (ft)

Figure 8-12 shows the relationship between the variables described in Equation 8-12. If the liquid level is below the centerline of the impeller, the static head is negative and is termed static suction lift. Table 8-4 provides vapor pressure heads for water at atmospheric pressure. In reality, the vapor pressure changes with air pressure, however, these values are reasonable for use with typical stormwater pump stations.

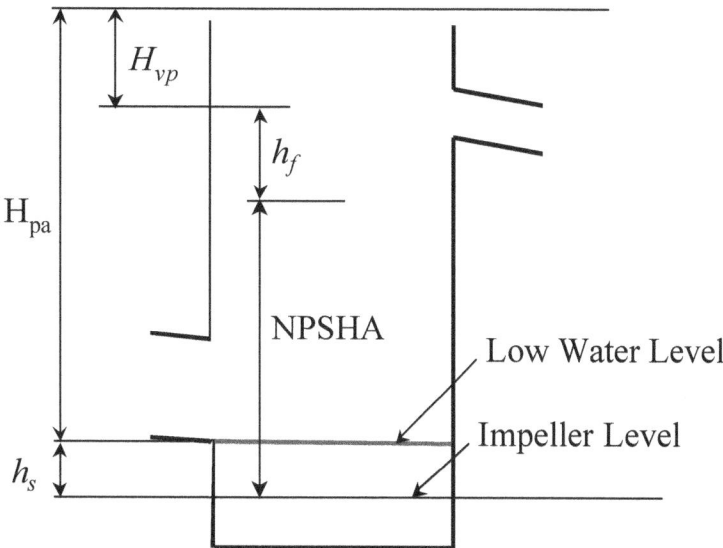

Figure 8-12. Net positive suction head available

Example: the lowest pumping level is 2.4 m (8 ft) above the proposed pump impeller, there is no suction line and bell losses are assumed to be negligible, atmospheric pressure is 10.3 m (33.9 ft) of water, and the vapor pressure head of water is 0.236 m (0.84 ft), the *NPSHA* is:

$$NPSHA = H_{pa} + H_s - h_f - H_{vp} = 10.3 + 2.4 - 0 - 0.236 = 12.46 \text{ m (40.9 ft)}$$

Table 8-4. Vapor pressure heads for water

Temperature		Vapor Pressure Head	
Celsius	Fahrenheit	m	ft
0	32	0.062	0.204
15	59	0.712	0.565
20	68	0.236	0.774
23.9	75	0.305	1.00
37.8	100	0.661	2.17

8.3.5 Net Positive Suction Head Margin

Net positive suction head margin is the amount by which the NPSHA exceeds the NPSHR. A positive margin is necessary; however, it is not practicable to define a recommended magnitude of the margin because of the many variables that affect the potential for vortexing and cavitation. The designer should check with the manufacturer of the selected pumps for recommendations of an appropriate margin based on the pumps and the conditions under which they will operate.

8.3.6 Submergence

Submergence is the static head of water required above the intake bowl or volute of the pump to prevent cavitation and vortexing. The designer must ensure that the minimum static head provides a NPSHA (see Section 8.3.4 - Net Positive Suction Head Available) that is an established margin (see Section 8.3.5 Net Positive Suction Margin) greater than the manufacturer's specified NPSHR (see Section 8.3.3 - Net Positive Suction Head Required). Also, refer to Section 9.1.2 - Sump Width for determination of required minimum submergence using Hydraulic Institute standards. The designer should use the criterion that creates the higher submergence.

8.3.7 Specific Speed

Specific speed, N_s, is often quoted as a dimensionless design index for pump impellers. However, strictly speaking, it has dimensions that are in terms of:

$[L^{0.75}T^{-0.5}]$ x revolutions per minute

where L represents units of length, and T represents units of time.

Specific speed is used to identify an upper limit on the shaft speed for any particular combination of total head, flow, and suction conditions. The specific speed is determined using the following equation:

$$N_s = C\frac{NQ^{0.5}}{H^{0.75}} \qquad (8\text{-}12)$$

where:

N	=	pump speed, rpm
Q	=	pump capacity at Best Efficiency Point (BEP), m³/hr (gpm)
H	=	pump head at BEP, m (ft)
C	=	unit factor = 0.861 * (1)

* For consistency with US manufacturers' pump data, a unit factor for metric units is given to result in a consistent value that is really in English units. The designer should verify the derivation and manufacturer's use of specific speed.

By applying pump affinity laws, a pump designer can predict the change in performance of a pump in response to changing speed. Discussion of affinity laws is beyond the scope of this manual. For a discussion, refer to a text such as *Pump Performance Characteristics and Applications.*[4]

8.3.8 Suction Specific Speed

Suction Specific Speed, N_{ss}, is a dimensionless index that is both representative of the geometry of the suction side of the impeller, and is used to determine the form and proportions of the impeller. The index can be used to select a pump maximum speed that will likely yield the smallest size of pump for the design conditions. It is determined using the following equation:

$$N_{ss} = C \frac{NQ^{0.5}}{NPSHR^{0.75}}$$ (8-13)

where:

$NPSHR$	=	net positive suction head required, m (ft)
N_{ss}	=	suction specific speed, rpm
N	=	pump speed, rpm
Q	=	pump capacity at Best Efficiency Point (BEP), m³/hr (gpm)
C	=	unit factor = 0.861 * (1)

8.3.9 Selection of Impeller Type

Table 8-5 indicates the ranges of specific speed over various impellers that are recommended for use.

Table 8-5. Specific speed range for impellers

Impeller Type	Specific Speed	
	Upper	Lower
Axial Flow	20,000	8,500
Mixed Flow (open)	10,000	5,000
Mixed Flow (closed)	6,200	4,000
Centrifugal	4,100	2,500

8.3.10 Pump Performance

Pump performance is measured as the variation in pumping capacity with respect to total dynamic head. Pump performance is a function of the following pump characteristics:

- pump type,
- pump size,
- impeller size, and
- speed.

Pump performance can be shown either as a single line curve for one impeller diameter or as multiple curves for the performance of several impeller diameters in one casing. Figure 8-13 shows a typical single pump performance curve.

Pump performance is measured and specified by the pump manufacturer. Most manufacturers' performance curves depict the total head developed by the pump, the brake power required to drive it, the derived efficiency and the net positive suction head required to drive it over a range of flows at constant impeller speed. Figure 8-14 shows a simplified version of a typical manufacturer's performance characteristics plot which presents performance curves for varying impeller size. Pump efficiency is also shown. The point at which maximum efficiency is achieved is represented by the best efficiency point, BEP, as shown in Figure 8-14.

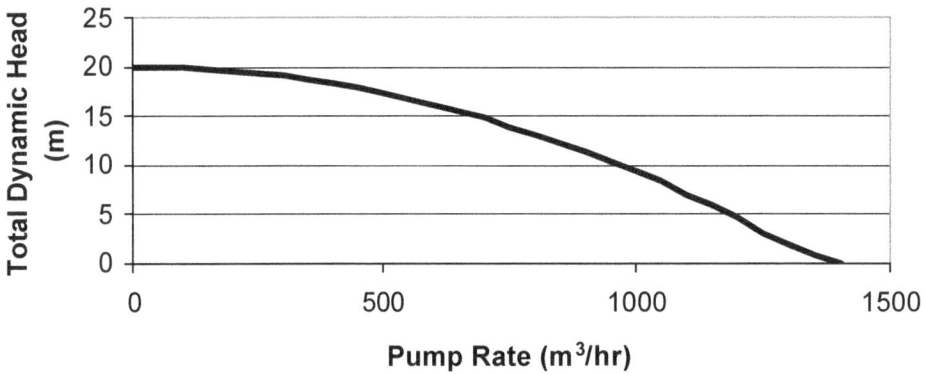

Figure 8-13. Typical pump performance curve

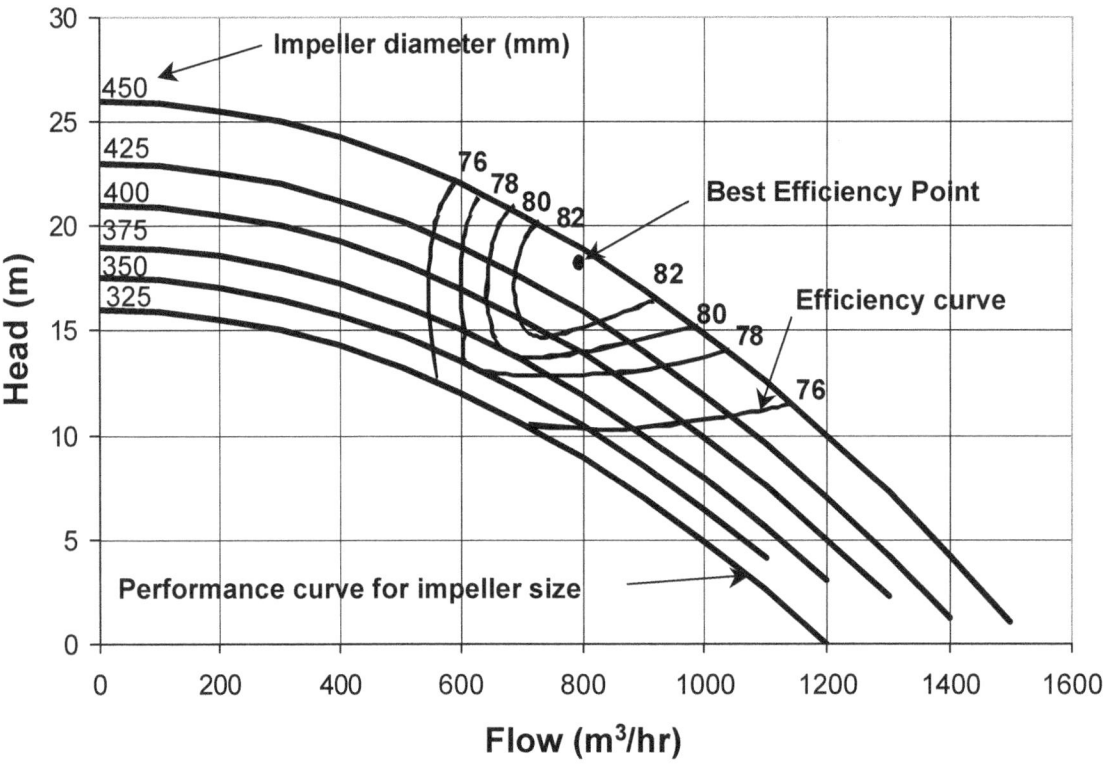

Figure 8-14. Simplified manufacturer's pump performance curves

8.3.11 Design Point and Operation Range

The design point is the target total dynamic head and discharge superimposed on the performance curve plot. The performance curve that is closest to the design point would be the characteristic curve for the desired pump. Alternatively, the manufacturer can usually trim the impeller to best meet the design point. If the efficiency at which the design point occurs is low, then the design should identify a different pump speed or pump type.

If a single pump were to operate with constant static head, the design point would be determined as the point of intersection of the system curve and selected performance curve, as shown in Figure 8-15. This situation is not applicable to highway stormwater pumps because the static head in a stormwater pump station varies during the operation of the pump station. The following two approaches are applicable to highway stormwater pumps depending on the configuration of the discharge piping.

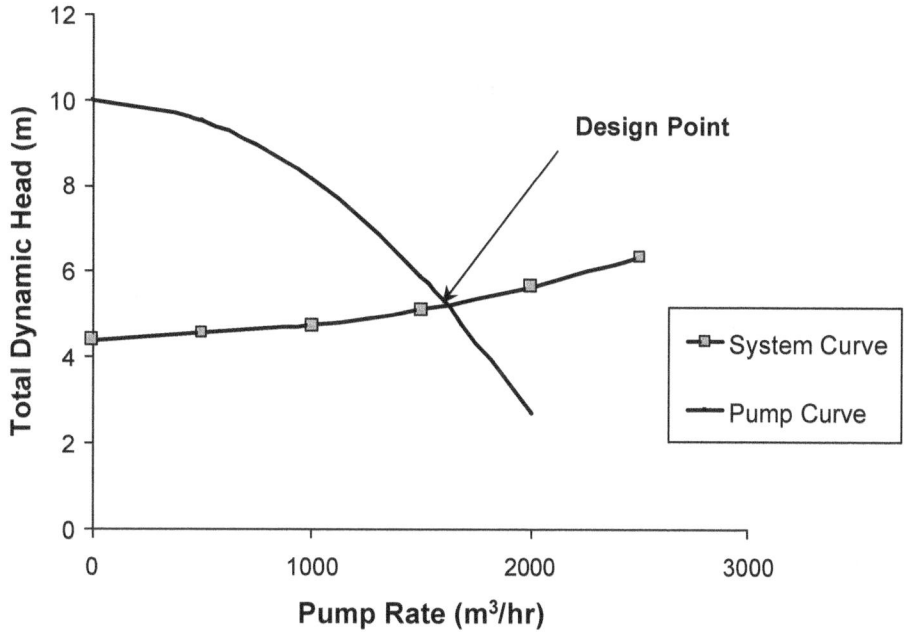

Figure 8-15. Design point

When two or more pumps connected to a common discharge line are operating and the static head changes over a limited range, the design point will move from A, for the first pump operating alone, to B, with both pumps operating at the minimum static head, as indicated in Figure 8-16. This represents an operating range. This case is typical for a stormwater pump station in which all pumps are connected to a common discharge line using a manifold.

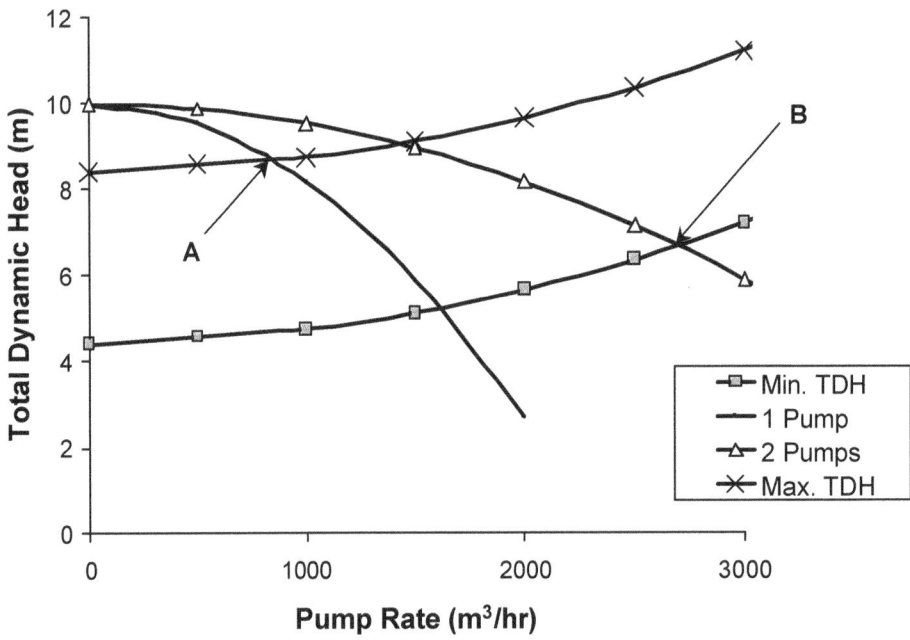

Figure 8-16. Operating range for two pumps and common discharge line

When two or more pumps with separate discharge lines are operating and the static head changes over a limited range, the range over which the pumps must operate is from A to B as shown in Figure 8-17. This case is typical for a stormwater pump station in which each pump discharges to the outfall via separate discharge lines.

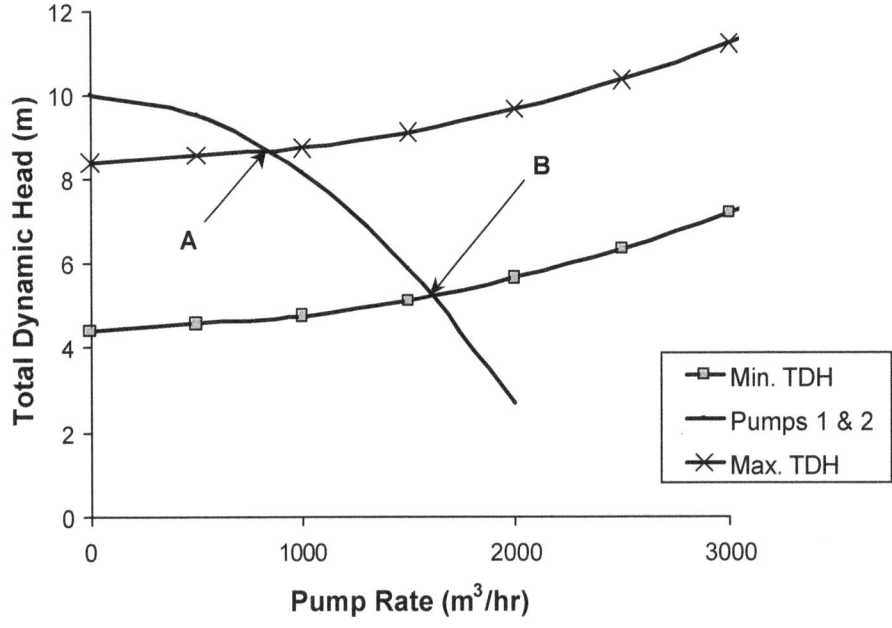

Figure 8-17. Operating range for pumps with separate discharge lines

8.3.12 Example of Pump Selection

An example is shown in Figure 8-18. The design pump rate is 800 m³/hr (7.9 cfs) at an average total dynamic head of 16 m (52.5 ft). The minimum, average and maximum total dynamic head curves (system curves) are superimposed on a simplified pump performance chart. The pump curve for a 425 mm (16.7 in.) impeller provides the design point (Point A). When the system operates at maximum total dynamic head, the actual pumping rate will be less at about 750 m³/hr (7.36 cfs) - Point B. When the system operates at minimum total dynamic head, the actual pumping rate will be greater at about 840 m³/hr (8.25 cfs) - Point C. Points A and B lie under the 40 kW (54 hp) power curve, but Point C lies between the 40 kW (54 hp) and 50 kW (67 hp) power curves. Thus, in order to ensure adequate power for the range of operations, a motor rated for 50 kW (67 hp) will be necessary.

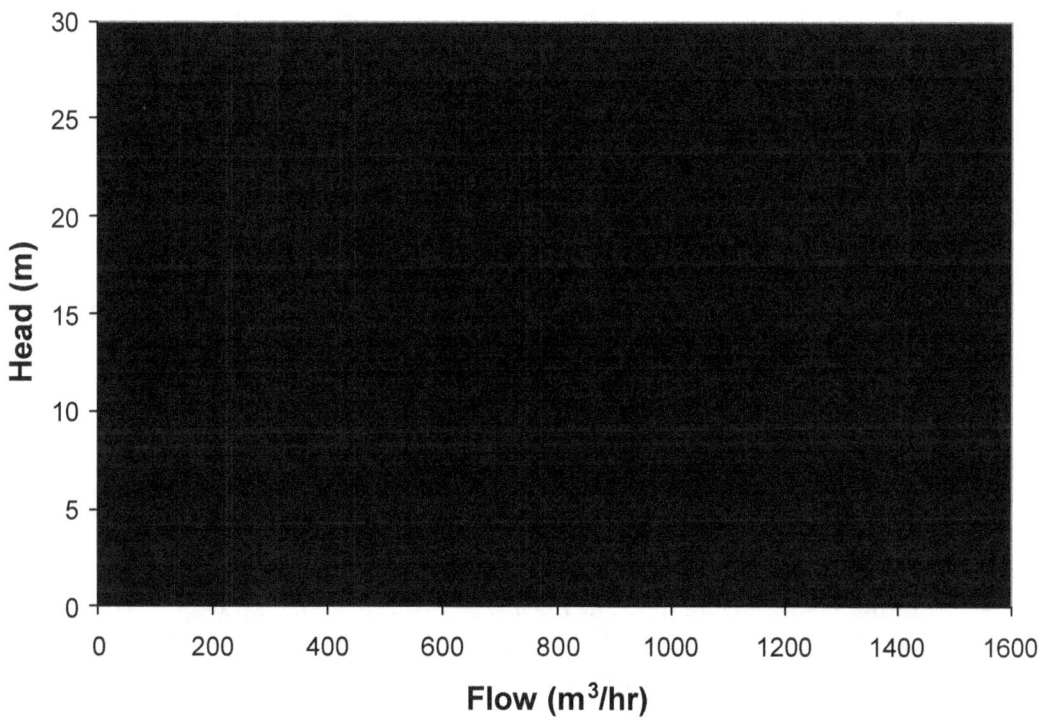

Figure 8-18. Manufacturer's performance curves

8.3.13 Procedure to Select Manufacturer's Pumps and Discharge Piping

You will need the following information to select a pump:

- desired pump type,
- manufacturer's catalogues of performance curves,
- design TDH_{ave}, TDH_{max}, and TDH_{min}, and
- design pump capacity.

Step 1.   Compute the Net Positive Suction Head Available using Equation 8-11.

Refer to the example in 8.3.4 Net Positive Suction Head Available.

Step 2.   Tabulate the following data:
- Approximate system head curve (Example: the TDH_{max}, TDH_{ave}, and TDH_{min} are 18.5 m (60.7 ft), 17.5 m (57.4 ft), and 16.5 m (54.1 ft), respectively.
- Design Capacity (Example: Individual pump capacity = 0.8 m^3/s (28.3cfs))
- Desired pump type (Example: try a submersible pump)

Step 3.   Browse manufacturer's catalogues for the family of pump performance curves that cover the desired pump capacity and operating range (TDH_{max} to TDH_{min}).

Step 4.   Select a performance curve that crosses through or near the design pump capacity at the design TDH_{ave}. This is the design point. If the manufacturer can trim the impellers to meet your exact design point, you can establish an intermediate performance curve by interpolation. If the design point falls on the performance curve at a low efficiency, you should preferably look for other pumps. The selection at this point is tentative.

Example: Figure 8-19 is a manufacturer's performance curve for a 715 rpm submersible pump. Assuming the manufacturer will trim the impeller, the design point is at a head of 17.50 m (57.4 ft) and discharge of 0.800 m^3/s (28.3 cfs). The required impeller size would be about 610 mm (24.0 in.) and the rated efficiency is about 85.7 %. Note that the motor power required would be 186 kW (250 hp).

Step 5.   For the selected pump curve, determine the operating discharges at TDH_{max} and TDH_{min}.

Example: Using Figure 8-19, at TDH_{max} = 18.5 m (60.7 ft), Q_{max} ~0.75 m^3/s (26.5 cfs) and at TDH_{min} = 16.5 m (54.1 ft), Q_{min} ~ 0.84 m^3/s (29.7 cfs). This represents the operating range for the selected pump.

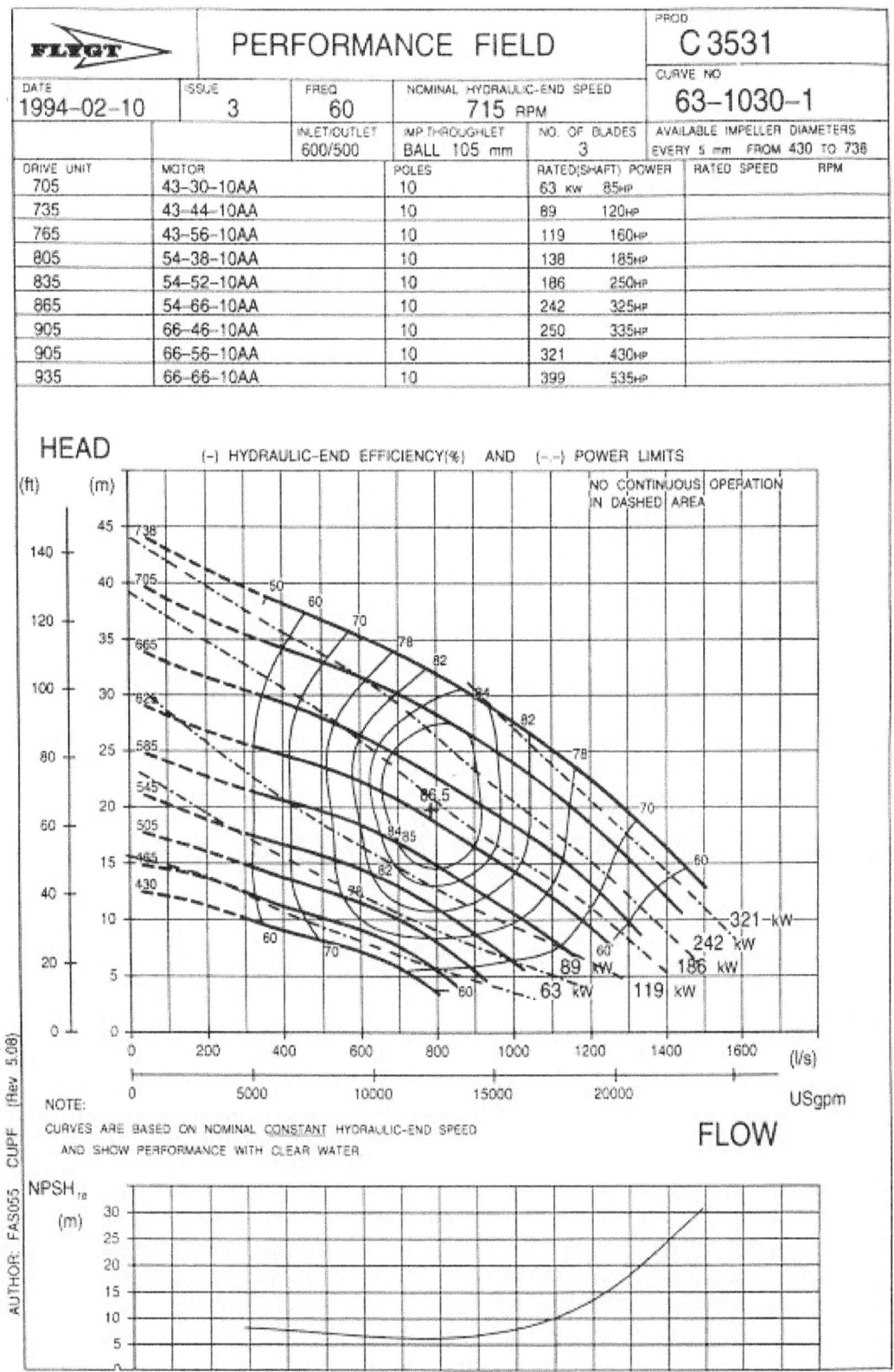

Figure 8-19. Pump manufacturer's curve

(Courtesy of Flygt Corporation)

8.3.14 Pump Bell/Intake Diameter

For design of highway stormwater pump stations, the recommended approach is to select the pump from manufacturer's curves as discussed in Section 8.3.13 - Procedure to Select Manufacturer's Pumps and Discharge Piping. Once the pump is selected, the bell or volute size can be identified from the manufacturer's specifications. The bell or intake volute diameter should then be checked to determine if the intake velocity falls within the recommended range indicated in Table 8-6.

Table 8-6. Recommended pump intake velocities

Adapted from reference 15

If the pump flow is:	The intake velocity can be:
< 0.315 m³/s (5000 gpm)	0.6 – 2.7 m/s (2.0 – 9.0 fps)
0.315 – 1.26 m³/s (5000 – 20,000 gpm)	0.9 – 2.4 m/s (3.0 – 8.0 fps)
> 1.26 m³/s (20,000 gpm)	1.2 – 2.1 m/s (4.0 – 7.0 fps)

If the intake velocity at the bell is higher than the higher limit established above, it will be necessary either to select a different pump, or add an umbrella to the pump. If the velocity is lower than the lower limit established above, the pump size may be excessive and the designer should consider selecting a pump with a small bell diameter, which, in turn, will reduce the required sump dimensions.

8.3.15 Power and Efficiency

Three terms are often used when referring to power for pumps:

- Water power
- Brake power
- Wire-to-water power

8.3.15.1 Water Power

Water Power WP (water horsepower WHP in English units) is the output power of a pump handling a given liquid at a given total dynamic head and discharge. It is determined using the following equation:

$$WP = \frac{1}{C_u}\gamma Q(TDH) \qquad (8\text{-}14)$$

where:

WP = water power, kW (hp)

C_u = unit conversion factor, 1000 (550)

γ = specific weight of liquid, N/m³ (lbf/ft³)

Q = discharge, m³/s (cfs)

TDH = total dynamic head, m (ft)

Refer to Table 8-7 for values of specific weight.

Table 8-7. Specific weight of water

Temperature		Specific Weight	
Fahrenheit	Celsius	English lbf/ft^3	Metric N/m^3
32	0	62.42	9805
41	5	62.43	9807
50	10	62.41	9804
59	15	62.37	9798
68	20	62.31	9789
77	25	62.24	9777
86	30	62.16	9764
90	32	62.11	9757
95	35	62.05	9747
104	40	61.94	9730
113	45	61.81	9710
122	50	61.68	9689

Example: The design pump rate is 0.2 m^3/s, the design TDH_{ave} is 11.06 m (36.3 ft), and the specific weight of water is 9798 N/m^3, the water power is:

WP_{ave} = 9798 x 0.2 x 11.06 / 1000 = 21.7 kW (29.1 hp)

Similarly at TDH_{max} =12.06 m (39.6 ft), WP_{max} = 23.7 kW (31.8 hp), and at TDH_{min} =10.06 m (33.0 ft), WP_{max} = 19.7 kW (26.4 hp)

8.3.15.2 Brake Power

Brake power, BP (brake horsepower, BHP, in English units) is the actual amount of power required to be supplied to the pump to maintain the water power. Brake power is the sum of the water power and power required to overcome losses incurred by the pump. Pump losses include:

- hydraulic losses,
- volumetric losses,
- mechanical losses, and
- disk friction.

Hydraulic losses incurred by the impeller and volute or diffuser are due to friction and momentum changes. Volumetric losses for centrifugal pumps are associated with leakage from the discharge side to suction side of the impeller. Mechanical losses result from friction between the moving parts of pumps such as between bearings, packing, and seals. Disk friction is the resistance between the impeller and casing.

The net effect of the above losses is described by efficiency that varies with pump type, TDH and flow rate. The efficiency is described as the ratio of pump power output to pump power input:

$$\eta = \frac{WP}{BP}$$

(8-15)

where:

η	=	pump efficiency
WP	=	water power, kW (hp)
BP	=	brake power, kW (hp)

The pump efficiency, η, is established by the manufacturer and is usually provided in the form of iso-efficiency lines on the pump performance curves such as the set shown in Figure 8-19.

> Example: For the manufacturer's pump selected in Section 8.3.13 - Procedure to Select Manufacturer's Pumps and Discharge Piping, the rated efficiency, η, is 75%, and using the water power computed in Section 8.3.15.1 - Water Power, the design brake power is:
> $$BP_{ave} = 21.7/0.75 = 28.9 \text{ kW (38.6 hp)}$$
> Similarly,
> $$BP_{max} = 31.6 \text{ kW (42.1 hp), and}$$
> $$BP_{min} = 26.3 \text{ kW (35.2 hp)}$$
> Thus, the selected pump requires a motor power of at least 31.6 kW (42.1 hp) to ensure enough power to handle the operating range. For this pump (Figure 8-19), the minimum motor power available is 67 kW (90 hp), which will suffice.

8.3.15.3 Wire-to-Water Power

In addition to the inefficiency of the pump, there is a loss of power associated with the motor. These losses are described by the motor efficiency, η_e, as the ratio of output power to the motor and input power to the motor. The wire-to-water power, then, is the total electrical power required to maintain the discharge and total dynamic head:

$$WWP = \frac{WP}{\eta \eta_e} \qquad (8\text{-}16)$$

where:

WWP	=	wire-to-water power, kW (hp)
WP	=	water power, kW (hp)
η	=	pump efficiency
η_e	=	motor efficiency

> Example: The maximum water power required is 23.7 kW (31.6 hp), the rated hydraulic efficiency is 75 % and the motor efficiency is 80%. The wire-to-water power is:
> $$WWP = 23.7/(0.75 \times 0.8) = 39.5 \text{ kW (52.7 hp)}$$

Some manufacturers may provide TDH, BP, and efficiency curves on a single plot for a single specific pump, impeller size, and speed. However, a majority of manufacturers present multiple performance curves, power curves and efficiency curves on a single plot, such as is shown in Figure 8-14. Generally, the iso-power lines represent only commercially available electric motor sizes. Also, the designer must note that the power curves relate to water at a specific gravity of 1. If the specific gravity of the liquid to be pumped is different, the power must be adjusted accordingly. Once a pump size is selected, it is usual to select a motor size based on the highest power curve within which the full performance curve of the selected pump size will operate. For electric motors, the designer may estimate the wire-to-water power by dividing the brake power by an assumed electrical efficiency of 0.8.

This page intentionally left blank.

9. SUMP DIMENSIONS AND SYSTEM CHECKS

9.1 SUMP SIZE AND CLEARANCE CRITERIA

The premise for the criteria is the pump bell diameter, D (Section 8.3.14 - Pump Bell/Intake Diameter). Generally, the most conservative of the manufacturer's specification for a specific pump and the Hydraulic Institute standards should be followed. The sump dimensions and clearances currently recommended by the Hydraulic Institute appear in Pump Intake Design[15]. The designer should refer to this document for detailed design. However, the appropriate criteria for basic stormwater pump station sumps appear here. The Hydraulic Institute recommends model testing for sump and pump configurations and conditions that deviate from their recommended criteria.

9.1.1 Sump Depth

The depth of the sump system is controlled by one or more of the following:

- invert elevation of the inlet
- typical wet well detail used by highway agency (location of trash racks etc.),
- volume required for cycling where no other significant storage is provided (See, Usable Storage),
- submergence requirements of the pumps used, and
- clearance required for the pump intake bell.

Hydraulic Institute recommendations[15] indicate that the invert elevation of the intake line (from the collection system or storage unit) should be no lower than the lowest pumping level. Since it is also desirable to maximize the storage within the storage system, the preference is to set the lowest pumping level at or slightly above the invert of the intake. The sump system floor should be set no lower than needed to accommodate the above conditions unless foundation requirements indicate otherwise.

9.1.2 Sump Width

The minimum total sump width is the sum of required pump bay widths and divider wall thickness for a rectangular wet well. Refer to Section 9.1.4 - Rectangular Sumps for required pump bay width. For a circular sump, the width is the diameter of the sump, which can be established by referring to Section 9.1.5 - Circular Sumps.

9.1.2.1 Sump Velocity

Water velocity in the sump should be low, preferably about 0.3 m/s (1.0 ft/s). The Hydraulic Institute[15] recommends a maximum velocity of 0.5 m/s (1.5 ft/s) for clear water. Several ways to reduce velocities include:

- increasing the inflow conduit size,
- using baffles,
- providing transition from inflow conduit to the wet well, and
- extending the length of the wet well.

The Hydraulic Institute[15] suggests attaining approach velocities of 1.0 m/s (3 ft/s) or more in the storage unit and pump bays of rectangular sumps containing sediment-laden water unless other measures are taken.

9.1.3 Submergence Depth

The minimum height of water above the pump intake bowl or volute, S, should be the higher of that required by the manufacturer for net positive suction head (See 8.3.6 Submergence) and the height determined using Equation 9-1, which is based on Hydraulic Institute standards[15] for pump bell velocities falling in the range shown in Table 8-6.

$$S = D + \frac{C_u Q_p}{D^{1.5}} \qquad (9-1)$$

where:

S	=	height from lowest pumping level to pump bell/volute, m (ft)
D	=	outside diameter of pump bell/volute, m (ft)
C_u	=	unit conversion coefficient = 0.935 (0.516)
Q_p	=	individual pump capacity, m³/s (cfs)

9.1.4 Rectangular Sumps

9.1.4.1 Limitations of Rectangular Sump Criteria

The Hydraulic Institute recommends use of the rectangular sump dimension criteria below the following pumping limits:

Total pumping capacity per wet well of 6.3 m³/s (223 cfs)
Individual pump capacity of 2.5 m³/s (89 cfs)

The Hydraulic Institute recommends that model tests be performed for stations that exceed the above limits. If the design exceeds these limits, then the designer should check to see if the pump manufacturer has performed tests to establish sump criteria. If not, the designer should do one of the following:

select another manufacturer's pumps with appropriate sump criteria,
use a larger number of smaller capacity pumps,
increase the storage unit size to reduce total pumping requirements, or
use multiple and separate wet wells.

9.1.4.2 Dimensions for Rectangular Sumps

Figure 9-1 identifies the recommended sump dimensions for a typical newly-designed rectangular sump. The dimensions apply to sumps used for wet-stations. Table 9-1 presents the Hydraulic Institute criteria for the sump dimensions. The dimensions apply to sumps used for wet-pit stations and to dry-pit stations with less than five diameters of suction piping. If fixture details, structure details or some other need results in a wider bay entrance, the width of the bay at the pump bell/volute, w, should be reduced to twice the bell/volute diameter as shown in Figure 9-2.

Table 9-1. Rectangular sump dimensions

Adapted from reference 15

Dimension	Description	Recommended Value
A	Distance from centerline of pump inlet bell/volute to sump entrance	5D
a	Length of constricted bay section at pump	2.5D
B	Clearance from back wall to centerline of pump inlet bell/volute	0.75D
C	Clearance between pump inlet bell and sump floor	0.3D - 0.5D
D	Outside diameter of pump inlet bell/volute	See 8.3.14 Pump Bell/Intake Diameter
H_{min}	Minimum water depth in sump	S + C
h	Minimum height of constricted bay section	Greater of H or 2.5D
S	Minimum pump inlet bell submergence	See Equation 9-1
W	Pump inlet bay width	$\geq 2D$
w	Minimum width of constricted bay section	2D
X	Pump inlet bay length	$\geq 5D$
Y	Distance from centerline of pump inlet bell/volute to screen	$\geq 4D$
Z_1	Distance from centerline of pump inlet bell/volute to diverging walls	$\geq 5D$
Z_2	Distance from centerline of pump inlet bell/volute to sloping floor	$\geq 5D$
α	Angle of floor slope	0 – 10 degrees
β	Angle of wall convergence	0 – 10 degrees
ϕ	Angle of convergence from constricted area to pump bay walls	10 degrees max.

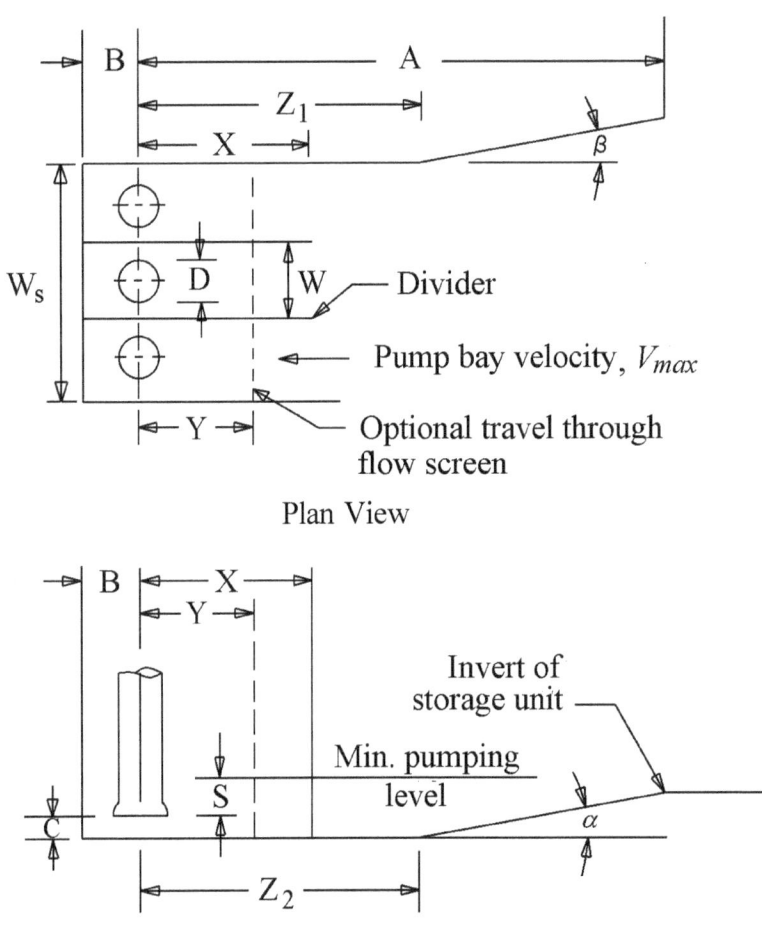

Figure 9-1. Recommended rectangular sump

Adapted from reference 15

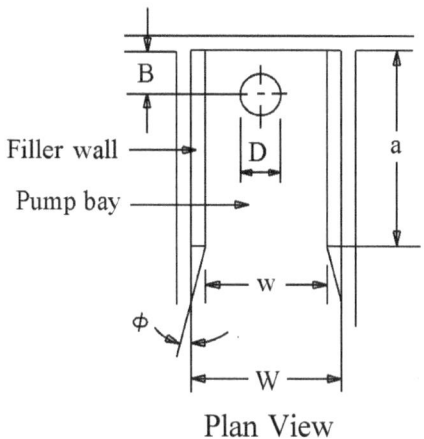

Plan View

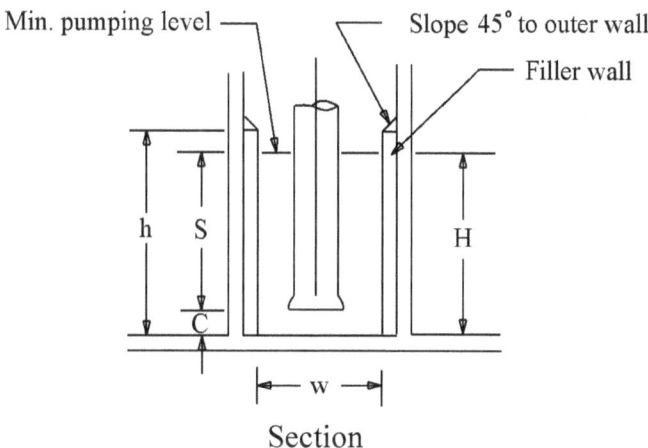

Section

Figure 9-2. Filler walls for pump bay

Adapted from reference 15

9.1.4.3   Contracting versus Expanding Sump Entrances

Figure 9-1 presents a contracting entrance in which an angle, β, is applied to converging walls. This would be appropriate when the inflow conduit is wider than the width of the sump.

When the inflow conduit is narrower than the sump width, the flow must expand when entering the sump. A transition with diverging walls of at least 10 degrees is recommended. Baffles may be used to help reduce velocities and distribute the flow.

9.1.4.4   Pump Bay Velocity

The velocity in the pump bay should be 0.5 m/s (1.5 fps) or less. The inflow velocity to a pump bay is determined using the following equation:

$$V_s = \frac{Q_p}{WH_{min}}$$

(9-2)

where:

V_s = average velocity in pump bay, m/s (fps)
Q_p = individual pump capacity, m³/s (cfs)
W = pump bay width, m (ft)
H_{min} = depth from the lowest pumping level to the sump floor, m (ft)

9.1.4.5 Example of Rectangular Sump Dimensions

It is necessary to establish sump dimensions and clearance values for a rectangular wet well in which will be placed two 0.2 m³/s (7 cfs) pumps with bell diameters of 600 mm (2 ft) each. The lowest pump stop elevation (minimum water level in the sump) is 16.750 m (54.95 ft). The pumps will be separated by a divider wall, the spacing of which is to be established. No screen is used inside the wet well. The inflow line is a 1.2 m diameter pipe with a clear and straight length of 160 m. The invert elevation of the inflow line at the sump is 16.75 m (54.95 ft).

The total pumping capacity is 2 x 0.2 = 0.4 m³/s (14 cfs)

The velocity at the bell face is:

$$V = Q / A = 0.2/(\pi \times 0.6^2 / 4) = 0.71 \text{ m/s (2.3 fps)}$$

Referring to Table 8-6, the flow is less than 0.315 m³/s and the velocity is between 0.6 and 2.7 m/s. Also, the individual pumping rate is less than 2.5 m³/s (89 cfs) and the total pumping rate is less than 6.3 m³/s (223 cfs). Thus the Hydraulic Institute recommended sump dimensions apply.

Using Equation 9-1, the required minimum submergence is:

$$S = D + \frac{C_u Q_p}{D^{1.5}} = 0.6 + 0.935 \times 0.2 / 0.6^{1.5} = 1.0 \text{ m (3.3 ft)}$$

The remaining appropriate variables appear in Table 9-2 based on the criteria in Table 9-1.

Table 9-2. Sample rectangular sump calculations

Dimension	Calculation	Magnitude, m (ft)
A	5 x 0.6	3 (9.8)
B	0.75 x 0.6	0.45 (1.5)
C	0.3 x 0.6	0.18 (0.6)
H_{min}	1.0 + 0.18	1.18 (3.9)
W	2 x 0.6	1.2 (3.9)
X	5 x 0.6	3 (9.8)
Z_1, Z_2	5 x 0.6	3 (9.8)

Assuming a divider wall thickness of 150 mm (6 in.), the internal sump width, W_s, would need to be:

$$W_s = 2 \times 1.2 + 0.15 = 2.55 \text{ m (8.4 ft)}$$

The preliminary elevation of the sump floor is set a height H_{min} below the lowest pumping elevation:

Sump floor elevation = 16.75 - 1.18 = 15.57 m (51.08 ft)

Note: it will be necessary to check the net positive suction head available (NPSHA) with the net positive suction head required (NPSHR). Referring to Figure 8-19, for a design discharge of 0.2 m^3/s (7 cfs) which is 200 liters/s (3193 gpm), the NPSHR is about 3 m (9.8 ft). This is lower than the NPSHA of 12.46 m (40.9 ft) calculated in Step 1 of Section 8.3.13 Procedure to Select Manufacturer's Pumps and Discharge Piping. Thus the proposed sump floor elevation of 15.57 m (51.08 ft) is satisfactory.

The height difference between the sump floor and inflow line is 1.18 m (3.9 ft). Using a floor slope of 10 degrees:

Required vertical transition length = 1.18 / tan (10) = 6.69 m (22 ft)

The centerline of the inflow line is to enter the sump at mid-width to ensure even flow distribution. If the inflow conduit is wider than the sump width, a horizontal transition is needed to contract the flow from the inflow line to the sump. Using a suggested horizontal transition (expansion) of 10 degrees:

Horizontal transition length = (2.55 – 1.2) / 2 / tan (10) = 3.83 m (12.56 ft)

Figure 9-3 shows the resulting sump dimensions.

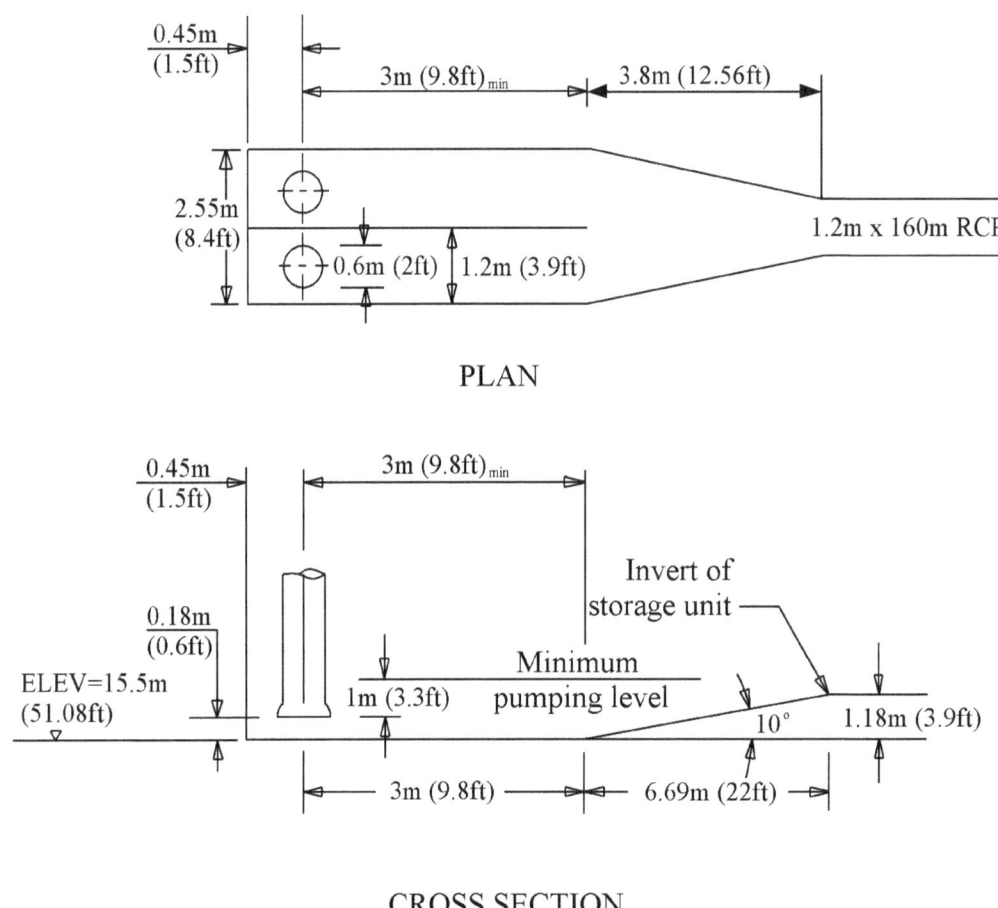

PLAN

CROSS SECTION

Figure 9-3. Dimensions for sump example (NTS)

9.1.4.6 Accommodation of Solids-bearing Water in Rectangular Wells

When rectangular wells are used and solids-bearing stormwater is anticipated, the preferred approach is to improve handling of sediments by creating confined pockets for the individual pumps and using steep slopes of 60 degrees or higher for vertical transitions to the confined pockets. The solids are then discharged by the main pump(s). Figure 9-4 shows a recommended rectangular wet well configuration[15]. Minimum slopes of 60 degrees are based on using concrete. The slopes could be reduced to 45 degrees if the walls are lined with PVC. The pocket depth should be a minimum of 2 pump bell diameters (2D). Individual pockets should be square with sides of 1.5 to 2.0 D length. The pump bowl or volute should be set D/4 above the floor of the wet well. This configuration requires a cone under each pump inlet (shown in Figure 9-4). For pump capacities in excess if 0.189 m³/s (3000 gpm) anti-rotation baffles are necessary.

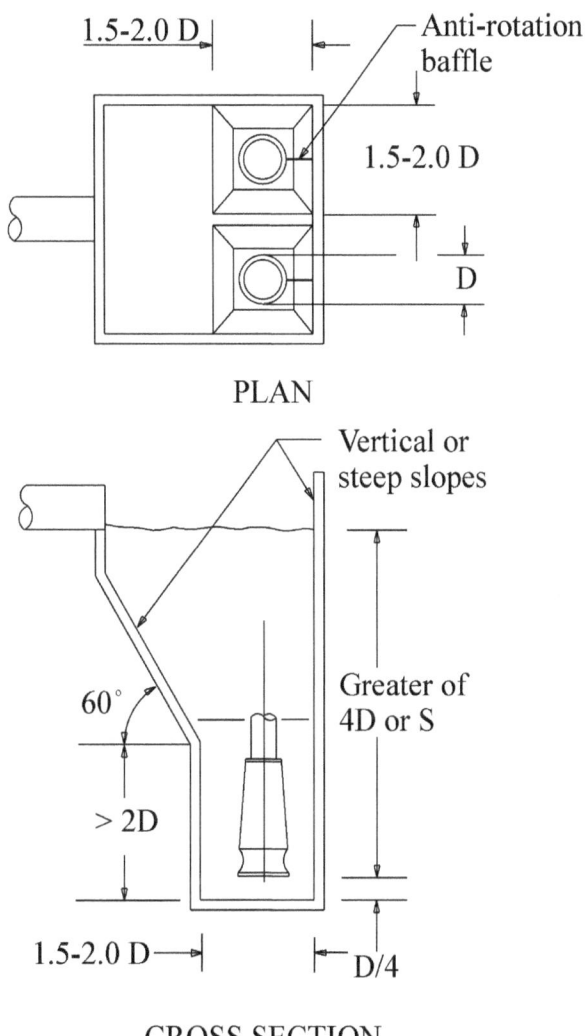

PLAN

CROSS SECTION

Figure 9-4. Confined rectangular wet well design

Adapted from reference 15

When using the wet well configuration identified in Figure 9-4, removal of settled solids will usually be effected when the pumps are activated. Floating solids could be removed by each pump if the water level is allowed to drop so low as to cause a strong vortex. However, each pump should be run separately. This may not be very practical for a stormwater pump because the well water level would have to be replenished after evacuation from the first pump. This would require an external source of water for flushing.

9.1.5 Circular Sumps

9.1.5.1 Limitations of Circular Sump Criteria

The Hydraulic Institute recommends use of the circular sump dimension criteria for individual pump capacities of 0.315 m³/s (5000 gpm) or less. The Hydraulic Institute recommends that model tests be performed for stations that exceed this limit. If the design exceeds this limit, the designer should check to see if the pump manufacturer has performed tests to establish sump criteria. If not, the designer should do one of the following:

select another manufacturer's pumps with appropriate sump criteria,
use a larger number of smaller capacity pumps,
increase the storage unit size to reduce total pumping requirements,
use a rectangular sump, or
use multiple and separate wet wells.

9.1.5.2 Dimensions for Circular Sumps

Figure 9-5, Figure 9-6, and Figure 9-7 identify the sump dimensions for various recommended configurations in a circular sump. The dimensions apply to sumps used for wet-pit stations. Table 9-3 presents the Hydraulic Institute criteria for the sump dimensions.

Table 9-3. Circular sump dimensions

Adapted from reference 15

Dimension	Description	Recommended Value
C	Clearance between pump inlet bell and sump floor	0.3D - 0.5D
D	Outside diameter of pump inlet bell/volute	See 8.3.14 Pump Bell/Intake Diameter
H_{min}	Minimum water depth in sump	S + C
S	Minimum pump inlet bell submergence	See Equation 9-1
C_b	Minimum clearance between adjacent bells/volutes	Greater of 0.25D or 100 mm (4 in)
C_w	Minimum clearance between bell/volute and closest sump wall	Greater of 0.25D or 100 mm (4 in)
D_s	Inside diameter of sump	See Table 9-4
D_p	Diameter of inflow conduit	Refer to 9.1.5.3 Inflow Pipe for Circular Sump
L_p	Unobstructed straight length of inflow conduit	$5D_p$

The following table identifies Hydraulic Institute recommendations for establishing the inside diameter of the sump, D_s. Refer to Table 9-3 for a description of variables. All dimensions should be in consistent units – i.e. m (ft). The Hydraulic Institute document, Pump Intake Design[15], identifies additional configurations.

Table 9-4. Recommended sump inside diameter (D$_s$)

Adapted from reference 15

Configuration	Reference Figure	Equation
Two pumps along sump centerline	Figure 9-5	$D_s = 2D + 2C_w + C_b$
Two pumps offset from sump centerline	Figure 9-6	$D_s = 2.5D + 2C_w + C_b$
Three pumps along sump centerline	Figure 9-7	$D_s = 3D + 2C_w + 2C_b$

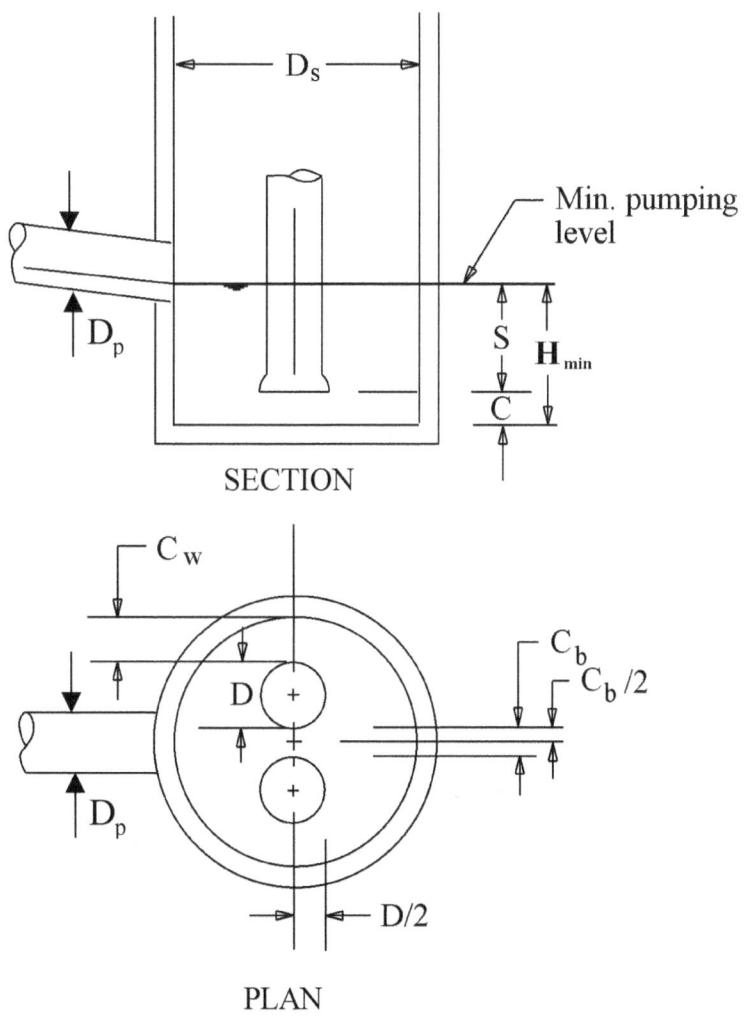

Figure 9-5. Circular sump – two pumps on center

Adapted from reference 15

SECTION

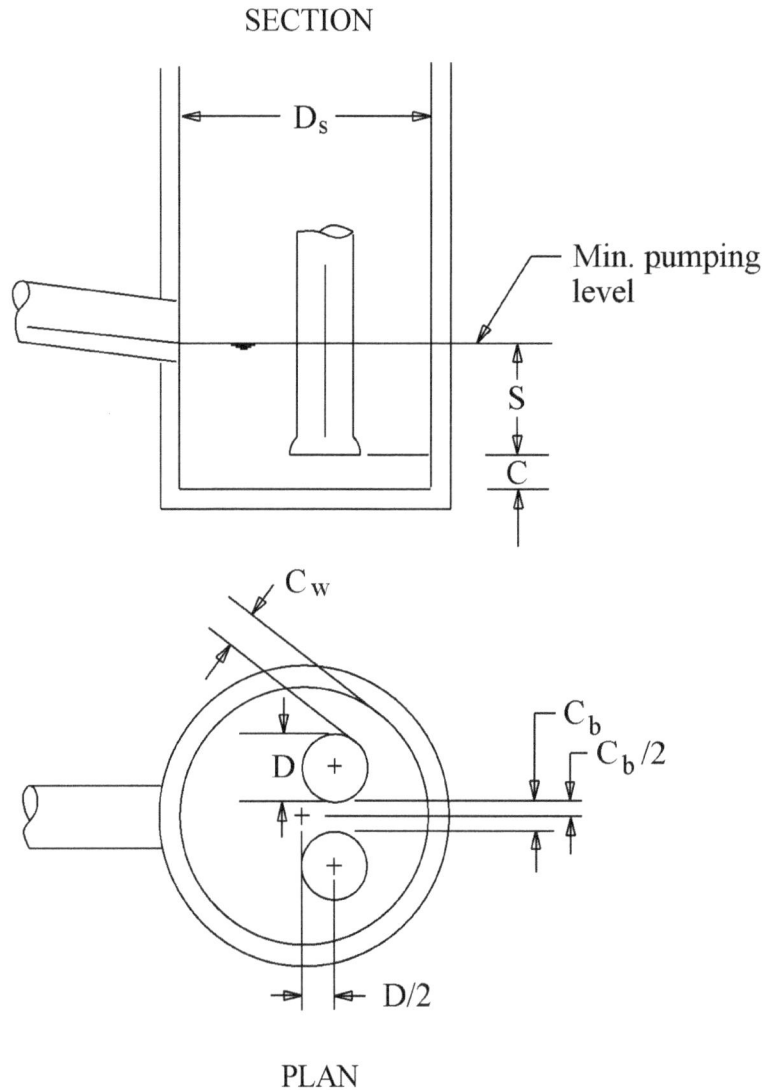

PLAN

Figure 9-6. Circular sump – two pumps off center

Adapted from reference 15

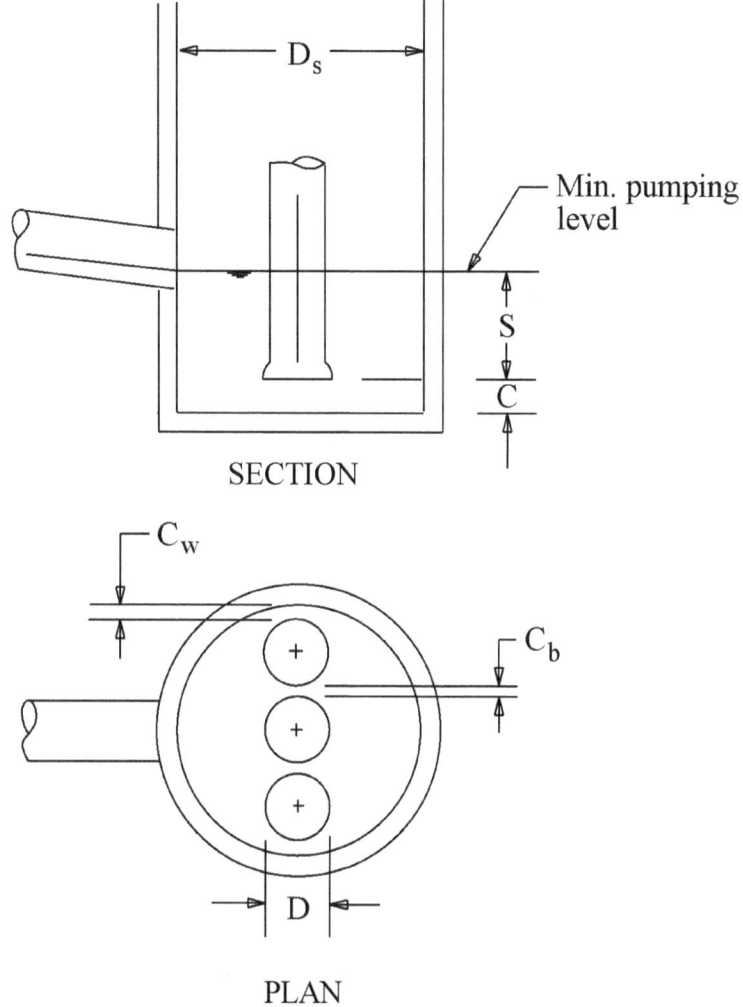

Figure 9-7. Circular sump – three pumps on center

Adapted from reference 15

9.1.5.3 Inflow Pipe for Circular Sump

It is recommended that the invert of the inflow conduit to the sump be set at the lowest pumping elevation (minimum water level) in the sump. This based on the following considerations:

1. The Hydraulic Institute[15] recommends that the inflow pipe invert not be higher than the lowest pumping level to minimize the potential for air entrainment and vortices resulting from free fall of the inflow.

2. If the invert of the inflow pipe is set below the lowest pumping level, the volume in the pipe below the pumping level will be ineffective with respect to available and usable storage.

The inflow pipe should be radial and perpendicular to the pumps to minimize rotational flow patterns. Also, the inflow pipe should be clear of any fixtures or obstacles for a distance of five pipe diameters upstream of the sump.[15]

The diameter of the pipe should be large enough to ensure that excessive inflow velocities are not incurred, yet small enough to minimize the potential for sedimentation. A practical inflow pipe velocity range appears to be about 0.6 m/s – 1.2 m/s (2 fps – 4 fps).

9.1.5.4 Example of Circular Sump Dimensions

It is necessary to establish sump dimensions and clearance values for a circular wet well in which will be placed two 0.2 m³/s (7 cfs) pumps with bell diameters of 600 mm (2 ft) each. The pumps will be offset from the centerline. The lowest pump stop elevation (minimum water level in the sump) is 16.750 m (54.95 ft). The inflow line is a 1.2 m diameter pipe with a clear and straight length of 160 m. The invert elevation of the inflow line at the sump is 16.75 m (54.95 ft).

The total pumping capacity is 2 x 0.2 = 0.4 m³/s (14 cfs)

The velocity at the bell face is:

$$V = Q/A = 0.2/(\pi \times 0.6^2 / 4) = 0.71 \text{ m/s (2.3 fps)}$$

Referring to Table 8-6, the flow per pump is less than 0.315 m³/s and the velocity is between 0.6 and 2.7 m/s. Also, the individual pumping rate is less than 2.5 m³/s (89 cfs) and the total pumping rate is less than 6.3 m³/s (223 cfs). Thus the Hydraulic Institute recommended sump dimensions apply.

Using Equation 9-1, the required minimum submergence is:

$$S = D + \frac{C_u Q_p}{D^{1.5}} = 0.6 + 0.935 \times 0.2 / 0.6^{1.5} = 1.0 \text{ m (3.3 ft)}$$

The remaining appropriate variables appear in the following table based on the criteria in Table 9-3.

Dimension	Calculation	Magnitude, m (ft)
C	0.3 x 0.6	0.18 (0.6)
H_{min}	1.0 + 0.18	1.18 (3.9)
C_b	0.25 x 0.6 = 0.15 (> 0.1 so OK)	0.15 (0.49)
C_w	0.25 x 0.6 = 0.15 (> 0.1 so OK)	0.15 (0.49)
D_s	See Table 9-4: (2.5 x 0.6) +(2 x 0.15) + 0.15	1.95 (6.4)
D_p	Given	1.2 (3.9)
L_p	5 x 1.2	6 (19.7)

9.1.5.5   Accommodation of Solids-bearing water in Circular Wells.

Circular wells can be designed or retrofitted to improve handling of sediments by minimizing the horizontal floor space in the sump except in the immediate area around the pump inlets and steep slopes for vertical transitions. A minimum slope of 60 degrees should be used for concrete or cement grout and 45 degrees for PVC-lined slopes. These measures increase the propensity for the solids to be suspended in the stormwater to be discharged by the pumps.

The basic well design should be as discussed in 9.1.5.2 Dimensions for Circular Sumps with modifications as shown in Figure 9-8.

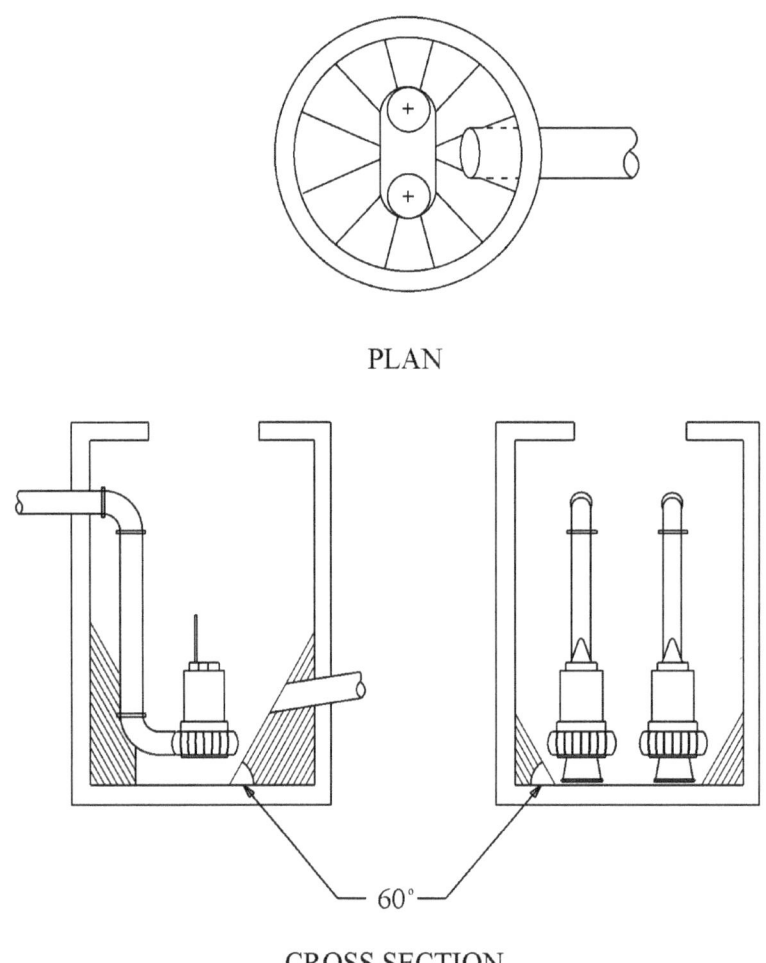

PLAN

CROSS SECTION

Figure 9-8. Circular well with sloping walls and minimized horizontal floor area

Adapted from reference 15

When using the wet well configuration identified in Figure 9-8, removal of settled solids will usually be effected when the first pump is activated. Floating solids can be removed by the pumps if the water level is allowed to drop so low as to cause a strong vortex. By necessity this will be lower than the required submergence, S, for normal operation and will cause vibration, noise, and high loads on the impeller. Thus, removal of floating debris should be limited to specific cleaning cycles that are applied for only short periods. The frequency and duration of such cleaning will vary from site to site and should be based on manufacturer's recommendations. The water level in the well should not be allowed to drop so low that the pump loses prime.

9.1.6   Wet Well for a Dry-Pit Station

Figure 9-9 defines the terms for the sump dimensions for the wet well of a dry-pit station. Refer to Table 9-1 for identification of the dimensions and recommended criteria.

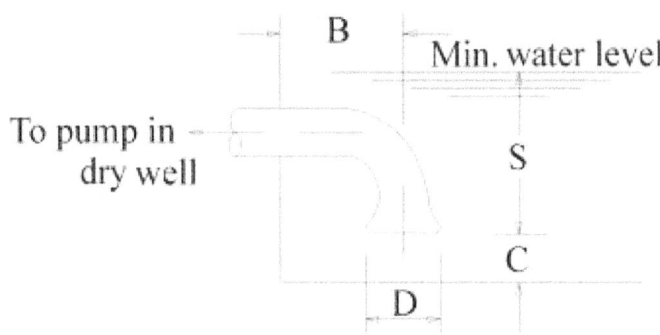

Figure 9-9. Sump dimensions for dry pit installation

Adapted from reference 15

9.1.7 Reduced Intake Velocity

If the pump bell/volute velocity is higher than the relevant limits established in Table 8-6, the use of an umbrella may be considered. However, providing an umbrella will not affect the net positive suction head requirement. A typical umbrella is shown in Figure 9-10. The diameter of the umbrella should be selected to provide an area that results in a recommended velocity of 1.7 m/s (5.5 fps) using the following equation.

$$D = C_u \left(\frac{Q}{V} \right)^{0.5}$$

(9-3)

where:

D	=	external diameter of umbrella, mm (in)
C_u	=	unit conversion factor, 1128 (13.54)
Q	=	pump capacity, m³/hr (gpm)
V	=	design velocity – recommended as 1.7 m/s (5.5 fps)

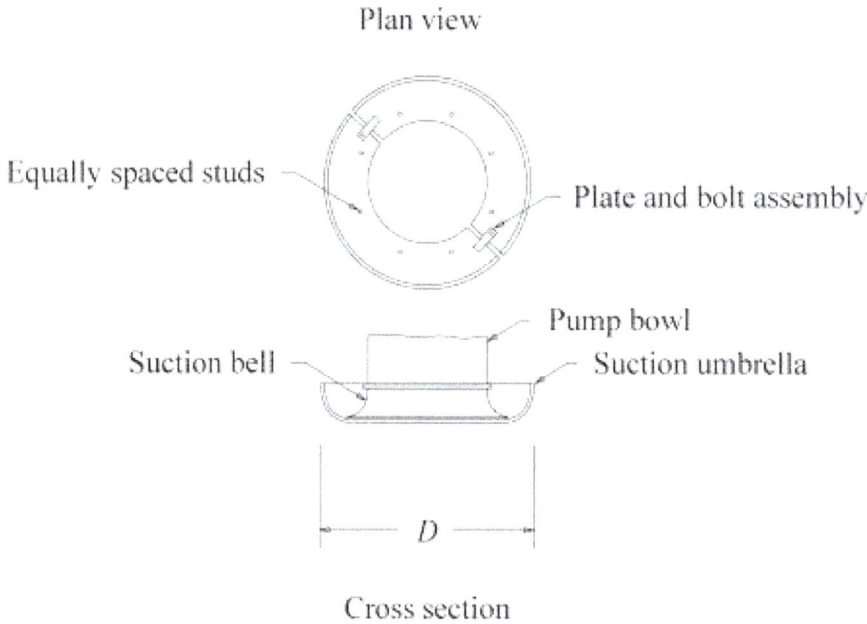

Figure 9-10. Typical umbrella

9.2 SYSTEM CHECKS

9.2.1 Check Storage Volume

The mass curve routing procedure results in a volume that must be stored for the selected pump capacities and operation levels. The designer must compare this value with the total volume available below the design highwater level. The total available storage provided must exceed the volume required. If not it will be necessary to make one or more of the following adjustments:

- Increase the size of the storage unit below the design highwater level.
- Increase the pump size.
- Adjust the start/stop elevations

In the example used in this manual, the volume determined from the mass curve routing procedure (Figure 7-6) was 223 m³ (7880 ft³). Referring to Table 6-2, the total available storage provided at a design stage of 2 m was 245.3 m³ (8668 ft³). Thus, sufficient storage was provided to avoid exceeding the maximum highwater in the wet well for the design runoff.

9.2.2 Check Cycling

If initial switching was based on approximate required cycle times, review the manufacturer's specified minimum cycle times. If the required times are greater than those used in Step 1 of Trial Switching you have several options:

- Select another pump with a lower required cycle time.
- Increase the storage to increase the cycling time for the established switching elevations.

- Adjust the switching elevations to increase the cycling storage.
- Employ an alternating switching scheme. For example, one in which the last pump off for one sequence becomes the last pump on for the next sequence. This is a recommended approach for design.

In the example used in this manual, the start and stop levels were set based on providing usable volume of at least 30 m^3 (Refer to 7.2.11 Procedure to Determine Trial Switching), the volume required for a cycle time of 10 minutes.

9.2.3   Check Frequency

Repeat the mass routing procedure using the selected check frequencies.

9.2.3.1   100-Year Flood

For the 100-year event, check the level to which the system is likely to flood. It is reasonable to expect the low point in the roadway to be inundated during the 100-year. However, you should ensure that the 100 year flood levels do not exceed any of the following:

- FEMA criteria, if applicable,
- state or local storm drain criteria for check floods, and
- critical elevations in wet well.

A critical elevation in the wet well would depend on the station type, but should be an elevation with freeboard, above which flood damage to any critical element of the pump system would be incurred.

9.2.3.2   Short Return Period

Checking the routing based on a runoff event of shorter return interval is often more useful than checking for possible short cycling. A two-year runoff of duration equal to the time of concentration would be an appropriate check.

9.3   BACKUP PUMPS

Conformance to the criteria established for minimum number of pumps reduces the risk of a complete failure but does not necessarily provide a safety margin for handling the design conditions. It is preferable to provide an additional pump over what would be required to meet the design total pumping capacity. Another way to provide backup capacity is to use significantly larger pumps than required. Generally, the designer should choose pumps that are slightly larger than required, but it is not advisable to oversize the pumps because of the higher potential for fast cycling. If the pumps are oversized, the designer should do a full evaluation of the performance of the system with the larger pumps.

This page intentionally left blank.

10. PUMP HOUSE FEATURES

10.1 INTRODUCTION

This chapter identifies and discusses major features of a pump house for a highway stormwater pump station. The pump house protects the pump control equipment, pump drivers, and ancillary equipment. The primary features are:

- ventilation,
- walkways,
- ladders or stairs,
- doors and hatches,
- cranes and hoists, and
- utilities.

10.2 PUMP HOUSE STRUCTURE

The pump house is usually rectangular and constructed with masonry. Figure 10-1 shows a typical pump house used by Texas Department of Public Transportation.

Figure 10-1. Typical pump house used by Texas DOT

10.3 VENTILATION

The need for ventilation is dependent on the pump station features. A list of ventilation considerations for use in pump station design follows:

- The wet well and dry well should be ventilated by an exhaust fan that draws air up from the bottom of the wet well in order to ensure a safe working environment for maintenance personnel.
- Where engine-driven pumps or generators are used, ventilation should be provided for fans in the pump-engine room that activate automatically when low-level vapor is detected.
- Stairwell ventilation system may consist of both an intake and an exhaust system, where each exhaust duct is equipped with a fire damper that closes automatically when smoke is detected in the engine room or wet well.
- Natural ventilation for the superstructure of the station can be provided by means of louvers and doorway openings. Louvers are more suitable than windows for pump station ventilation and louvers also discourage vandalism. Cross-ventilation is necessary for both electric motors and gas engines since the former generate heat and the latter require air for combustion. Fixed or manually controlled louvers suffice for electric motors, but engines should preferably blow open upon activation of the engines or have thermostatically controlled exhaust louvers.
- Submersible pumps do not usually require ventilation for cooling because they are cooled by storm water. However, air quality requirements will usually warrant suitable ventilation for the wet well and pump house.

Figure 10-1 shows cooling louvers and exhaust vents for diesel generators. Figure 10-2 shows louvers used for exhaust and cooling on a pump house used by Arizona DOT.

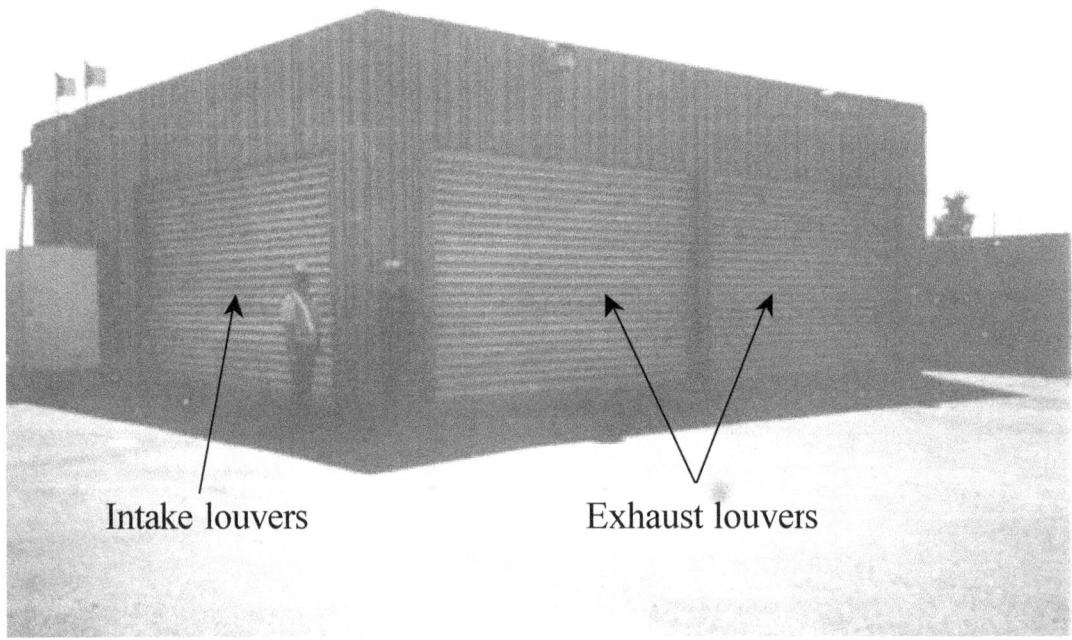

Intake louvers   Exhaust louvers

Figure 10-2. Ventilation on Arizona DOT pump house

10.4 WALKWAYS

Catwalks or walkways should be provided where necessary to ensure safe access to station components.

10.5 STAIRWAYS AND LADDERS

Stairways are preferred to ladders, since ladders are generally more hazardous, especially when tools or equipment are being carried.

If provided, the stairway from the pump pit to the engine room or pump room can be enclosed and a bulkhead provided with a gas-tight door that is fitted with a gasket. The gasket seals off the pump pit and prevents gasoline or other explosive vapors from entering the superstructure. Marine-type bulkhead doors are sometimes used, but their high threshold, restricted opening and screw clamps are inconvenient.

10.6 DOORS AND HATCHES

Personnel doors should be provided at each end of the pump room for easy exit. A wide roll-up door may be used for equipment and truck access. Access hatches to the pump pits, when open, must be protected by posts and surrounding chains. Similar provisions for guard posts and chains should be made at all other removable cover plates.

10.7 CRANES AND HOISTS

10.7.1 Overhead Cranes

Use of large pumps may warrant installation of a permanent lifting device such as an overhead crane, monorails, jibs or hoists. Figure 10-3 shows a permanent overhead crane installation with submersible pumps used by Arizona DOT.

Overhead cranes have three directions of movement that must be considered in the design of the station:

- longitudinal, for the bridge,
- transversely, for the trolley, and
- vertically, for the hoist.

The top-riding bridge-cranes with the end carriages on top of rails set on longitudinal girders are preferred to the underhanging type of crane, for which the bridge is hung from the longitudinal beams set below the roof slab or elsewhere in the structure. The type of crane selected, however, is primarily dependent upon the ability of the crane to handle the largest and heaviest predicted load.

Figure 10-3. Overhead crane over wet well with submersible pumps

10.7.2 Monorails

The travel path of the hoist is limited to one plane only along a monorail. The monorail system may be designed to accommodate any portion of the total weight of equipment with suitable safety factors.

The hoist may travel on a monorail by a hand racking system or may be electrically powered. The hoist should be electric motor operated by a pendant control or by a push button station located at a central location within the building. Power to the hoist is usually provided by a reeling cable or tag line. The hook should be mounted in a thrust bearing so that it may be rotated under rated load. The hoist should have a right and left hand grooved drum to avoid twisting of the block.

10.7.3 Jib-Cranes

Jib-cranes have limited use for handling mechanical equipment in a pump station. They have a horizontal arm pivoted at one end to allow it to swing. A hoist, electric or manually-powered, can be positioned as required along the horizontal arm. Jib-cranes are of most value in trash handling.

10.7.4 Hoists

Most hoists used in pumping stations are powered electrically for the lifting operation and may be suspended from trolleys that are then either manually or electrically operated.

Hoists should be provided in all stations for the handling of equipment and materials that cannot be lifted or removed easily from the station by manual labor. Hoists should be placed over basket screens, submersible pumps, other pumps or motors and other locations where it is necessary to lift heavy pieces of machinery or equipment. Where a pump pit is very deep it may be necessary to have the hoist fitted with cable drums which are larger than standard, in order to accommodate the extra length of cable.

The hoisting anchors are usually embedded in walls or overhead concrete, which enables rigging to be secured any time dismantling or removal of pumps is required.

10.8 OSHA REQUIREMENTS

The Occupational Safety and Health Administration (OSHA) establishes requirements for features such as those described above. All aspects of pump station design must comply with current OSHA criteria. Additionally, there may be stricter state or local criteria that should be considered.

10.9 FIRE PRECAUTIONS

10.9.1 Small, Unattended Stations

Small pump station houses constructed entirely with non-combustible materials and having small electrically powered pumps with means of immediate egress generally do not require special fire protection equipment. Not even a chemical fire extinguisher is required since any electrical fire would probably occur while the station was unattended. Although this would be damaging to equipment, it would not be hazardous to personnel.

10.9.2 Large Stations

The pumping equipment at larger pumping stations is typically more complex, and lubricating oil may be spilled. In such cases, chemical fire extinguishers should be provided and there should be doors at opposite ends of the building, equipped with panic bars. All fire protection provisions should be in accordance with state and local fire ordinances.

10.10 SECURITY AND SAFETY

The security and safety features of the pump house should maintain a secure and safe working environment by paying particular attention to the safety features involved in the operation and maintenance of the station. For a list of pump station safety features, refer to Chapter 12 - Construction, Operation and Maintenance.

This page intentionally left blank.

11. BASIC ELECTRICAL AND MECHANICAL INFORMATION

11.1 INTRODUCTION

The intent of this manual is to provide detailed guidance for the hydraulic design of pump stations. This chapter provides familiarity-level discussion for civil engineers so that the civil engineer can be make accommodations for other pump station needs and be conversant with engineers of other disciplines.

11.2 POWER SUPPLY

Requirements for pump stations include the power necessary to operate the pumps and the power required to operate other station components such as the pump switching, water level sensing, and lighting. Common power sources for stormwater pump stations include:

- public utility electricity,
- electric generators powered by natural gas and diesel, etc., and
- diesel, natural gas, or liquid propane engines.

11.2.1 Electrical Mains

11.2.1.1 Supply Lines

Electrical supply from a public utility company is the most common primary source of power for a stormwater pump station. The reasons underground electric service is preferred include:

- the reduced risk resulting from weather-related damage,
- lower visual impact, and
- reduction of potential hazards to cranes and other tall mobile equipment.

Even if the supply power lines are overhead, it is often practicable and desirable to tap into the lines at the station right-of-way boundary and provide underground conduit to the station.

If overhead lines are the only practicable means of providing power to the station, it is imperative that the power lines be kept clear of the driveway, station loading area, and turnaround area. Additionally, clear warning signs should be posted.

11.2.1.2 Electricity Rates

The quality of electrical service will depend on how well the system maintains a constant voltage under varying load demands and its susceptibility to outage, especially during storms. Therefore, a secondary source of electricity is desirable. Where practicable, this can be an alternate mains supply.

Utility companies are now deregulated and competition may affect rate schedules. The rates may also vary with a wide array of factors including:

- load demand,
- energy consumption,
- power factor,
- available voltages,
- primary service cost,
- transformer and substation ownership,
- fuel cost, and
- demand interval.

Estimates of the load data and potential demand for initial and future requirements of the pump station will be needed to evaluate potential costs.

11.2.2 Power from Generators

In remote areas, generators may be the primary source of power. When generators and engines are used, several fuel options exist. Natural gas can be piped underground to the station by a public utility, but such service is usually available only in urban areas. Liquid propane gas and diesel can be stored on site, such as is shown in Figure 11-1. Proper access for supply vehicles must be maintained, with all locations of storage above any flood hazard. The choice of engine type and fuel type is usually based on power requirements, availability and economics. In EPA non-attainment areas for air quality, use of natural gas may be preferable to diesel. Fuel storage areas must be designed to contain spills at the site.

Figure 11-1. Liquid propane tanks for pump station power

11.2.3 Power Use

If electric motors are used the following table provides recommendations.

Table 11-1. Recommendations for power use

Item	Preferred Criterion
Motor Type	Constant speed, 3-phase induction
Motor Voltage	440 V - 575 V
Maximum Power Rating	225 kW (300 hp)
Max Power when using portable backup power	56 kW (75 hp)

11.2.4 Electrical Power Factor

When an AC electric motor is running, the cycles of voltage and current may be out of phase with each other, so that the product of voltage and current, called apparent power, will be greater than the actual power consumed by the circuit. The ratio of actual power consumed by a circuit to the apparent power of that circuit is called the power factor. Power factor is expressed as a number usually between 0.8 and 1.0. As high a power factor as possible is desired because this is one of the items considered by the serving utility in determining its billing to the user. There is a cost penalty for low power factor. Capacitors are sometimes used in a circuit to improve power factor.

11.3 MOTORS

Electric motors account for a high percentage of all pump drivers. Figure 11-2 shows a typical motor for a vertical pump. Alternating current (AC) electric motors are the most frequently used drivers in pumping stations primarily because of their versatility, compactness, low maintenance and reliability.

Figure 11-2. Typical electric motors for vertical pumps

The three most common types of AC electric motors used to drive pumps are:[1]

squirrel-cage induction,
synchronous, and
wound rotor-motors.

Squirrel-cage induction motors are by far the most common motors used to drive pumps and generally operate at a single speed. Synchronous motors are similar to squirrel-cage induction motors except that they operate at speeds which are determined by the supply electrical frequency. Synchronous motors are used for pump applications requiring larger horsepower ratings at lower speed conditions. Wound-rotor induction motors are similar to squirrel-cage but can be used either where torque control is required or where variable speed is necessary. Variable speed motors are not preferred for use with highway stormwater pumps.

11.3.1 Other Motors

Other AC motors include the brushless synchronous motor; the water-cooled motor, which is typically quieter than air-cooled motors; and the low inrush current motor, which, along with the water-cooled motor, is generally preferred for pumping applications requiring higher power.

Direct current (DC) motors are only occasionally used in pumping stations and are used only when DC is the only power available.

11.3.2 Service Factor

Service factor is essentially a safety margin that refers to the motor's ability to continuously deliver horsepower beyond its nameplate rating under specified conditions. Motors are designed with an allowable rise in temperature above ambient temperature during operation. The maximum allowable temperature rise during operation for a motor varies with respect to insulation class and the motor's service factor. Most motors are rated with a 1.0 or 1.15 service factor. A 10 kW (13.3 hp) motor operating under rated conditions with a 1.15 service factor should be able to continuously deliver 11.5 kW (15.3 hp) without exceeding the NEMA allowable temperature rise for its insulation system. NEMA allows an ambient temperature of 40 C (104 F) when specifying "usual service conditions."

If the ambient temperature exceeds 40 C or at elevations above 1005 m (3,300 feet), the motor service factor must be reduced or a higher horsepower motor is required. As the oversized motor will be underloaded, the operating temperature rise is less and overheating will be reduced.

11.3.3 Motor Starting

During starting, the electric motor draws a high current that can be as much as ten times the normal operating current. The current surge incurs voltage drops and electromagnetic interference that can affect the electrical distribution system. Power companies or municipal agencies usually establish limits on the magnitude of current surges. Several wiring methods, such as the following, may be used to start the motors:

- across the line
- automatic transformer
- wye-delta
- electronic

"Across the line" does not provide voltage control. The latter three systems control current surges by reducing the starting voltage. The reduced starting voltage is accompanied by reduced starting torque for the motors. The electrical designer must ensure that the starting voltage provides enough torque to the desired starting motors.

11.3.3.1 Across the line

If the motors for pumps require a lower surge current than specified limits, the normal "across the line" starting method can be used in which the motors are directly wired to the power distribution system.

11.3.3.2 Automatic Transformer

One of the most common and preferred starting methods is the use of three automatic transformers which reduce starting voltage for a specific period of time then revert control to across the line. Typical automatic transformers allow connections for voltage reductions of 50%, 60%, and 80%.

11.3.3.3 Wye-delta

This is a complex connection that is usually used with specific motors that have six-lead wiring. Typically, there is a fixed voltage reduction of about 58%.

11.3.3.4 Electronic

Electronic switches known as thyristors can be set to vary the voltage gradually during starting to ensure that the current surge does not exceed a specified limit. Though, seemingly they would be preferred because of the degree of control, they are usually more expensive than other starting methods.

11.3.3.5 Implications of Starting Methods

For the hydraulic designer, it may be necessary to consider the length of time it is likely to take for the pumps to reach operating speed and to check that there is sufficient storage available to accommodate the additional inflow during start-up time. However, if the appropriate cycling volume is provided, potential lag times should not usually be a significant issue.

11.3.4 Advantages of Using Electric Motors

The advantages of using electric motors for driving stormwater pumps are as follows:

- they are compact and have minimum space requirements,
- installation is usually less expensive than for engines,
- they are generally easy to remove for maintenance or replacement,
- they are available in a variety of sizes,
- various insulation systems are available, and
- numerous mounting and enclosure configurations are available.

11.3.5 Disadvantages of Using Electric Motors

The disadvantages of using electric motors for driving stormwater pumps are as follows:

- they are susceptible to damage by flooding,
- high starting currents are required for most types,
- they are subject to utility outages,
- an alternate power source will be required for the system,
- space is required for control equipment,
- expensive and complex control systems are required for some types,
- they are subject to overheating, and
- they may be more expensive to operate than engines.

11.4 ENGINES

Engines are used only sparingly as pump drivers in small pump stations. Engines are usually considered inferior to electric motors for these stations because of the noise and vibrations they produce, the space they occupy, and the maintenance and operating supervision required. Figure 11-3 shows an engine that drives a vertical pump using a right angle gear. The use of engines should be considered if the power supply is unreliable or expensive and the environmental problems appear insurmountable. Engines are used more commonly in large pump stations.

Figure 11-3. Engine drive

11.4.1 Engine Categories

Engines that might be used to drive pumps in pumping stations can be grouped according to three characteristics:

ignition,
cycle, and
configuration.

Engines based upon ignition can be subdivided into those using spark ignition and compression ignition. Engines using spark ignition require a spark to ignite the fuel. In compression ignition engines, the heat generated by the compression of the fuel-air mixture ignites the fuel.

Two and four-stroke cycle designs are available for stationary engines. The two-stroke cycle engine generally is limited to compression ignition designs and is less efficient, more complex, lighter in weight, less expensive, and noisier than four-stroke cycle engines. Four-stroke cycle engines are available as both spark ignition and compression ignition types.

In-line and V-block configurations are available for stationary engines. V-type engines are generally more expensive to maintain, smoother operating, more compact, and require less floor space than in-line engines.

11.4.2 Advantages of Using Engines

The advantages of selecting an engine over an electric motor include:

- greater reliability,
- economy of operation,
- improved protection against surge damage, and
- the ability to be placed in remote locations or where electrical supply is unreliable.

11.4.3 Disadvantages of Using Engines

The disadvantages of selecting an engine over an electric motor include:

- concern over noise emissions,
- concern over air pollutant emissions,
- lack of adequate maintenance capability, and
- high capital cost.

11.5 BACKUP POWER

Where electrically driven pumps are used, an alternate source of electricity is essential unless an inoperative pump station can be tolerated during a storm. The choice of backup is usually based on economics.

11.5.1 Alternate Sources of Electrical Power

Alternate sources of electrical power, in case the service from the utility company is interrupted, include:

- a dual power supply with automatic transfer,
- a dual power supply with interlocking circuit breakers,
- permanent back-up generators, and
- mobile generator, manual transfer switch, and standby power (SBP) receptacle.

Figure 11-4 shows a diesel-powered generator used by Texas DOT as a standard feature of their stations in Houston. Primary power is at high risk for interruption during the frequent thunderstorms that are experienced in Houston. There is usually a short time delay while the backup power is engaged. Preferably, there should be a safety margin of storage volume in the system to accommodate the design runoff volume that would occur during the time delay.

Figure 11-4. Diesel-powered backup generator

The maximum power rating of available mobile generators will limit the power rating for the pump motors. Figure 11-5 shows a mobile generator and SBP used by Caltrans. Caltrans limits the main pump ratings to about 56 kW (75 HP) to be compatible with the mobile generators. If the local preference is to use mobile generators, the designer should consider the average response time for providing the mobile power. The storage system should be sized such that the design volume of rainfall during the response time interval can be held without exceeding the maximum highwater elevation.

Figure 11-5. Mobile generator and standby power receptacle

11.5.2 Backup Engines

Dual-drive pumps are a viable alternative for most dry-pit stations and some wet-pit stations that use vertical pumps. The electric motor is usually the primary driver and is placed on the main axis of the pump drive shaft. The engine drives a shaft that is connected to the main shaft via a right angle gear drive and clutch.

11.6 ELECTRICAL SYSTEMS

Primary electrical systems that should be considered in the design of a pump station include systems that:

- provide surge protection for pump station equipment during electrical service irregularities or failures from the utility company,
- provides for phase failure,
- provides protection against lightening strikes,
- provide automatic power transfer, and
- allow the standby engines to start automatically in response to sensing a power outage and a change in water level in the wet well, since most pumping stations are unattended.

11.6.1 Protection Systems

A switching system should minimize damage to the pumps and limit the extent and duration of interruptions in the electrical service supplied to the station. A system must isolate any affected portion of a station system while maintaining normal service for the rest of the system. It must minimize the danger of short-circuit currents and may provide alternate circuits to minimize the duration and extent of outages.

System protection is provided by the following devices:

- a fuse, which is the simplest of all protective devices, performing both sensing and interrupting functions.
- circuit breakers, which are interrupting devices and must be used in conjunction with sensing devices to fulfill the detection function. Most low-voltage applications use either molded-case circuit breakers or other low-voltage circuit breakers having series sensing devices built into the equipment.

Circuit breakers and fuses are employed to disconnect the affected parts of the power system. The use of computer-controlled circuitry to perform the sensing and timing functions can help optimize control conditions and allow for easier testing, monitoring and adjustments.

When two utility sources from separate supply sources and over separate lines are connected, an automatic transfer switch is usually located in the main distribution equipment. Special consideration should be given to the ability of the switch to:

- close against high inrush currents,
- carry full rated current continuously from normal and secondary source,
- withstand fault currents without contact separation, and
- have adequate interrupting capacity.

11.6.2 Engine and Motor Switches

Since most pumping stations are unattended, engines must start automatically. The engines are designed to start at a predetermined low speed or load. After a suitable time delay the throttle is opened to its operating speed. A selector switch is included on the control panel.

During a power outage, a transfer switch should be provided with an accessory control with switches that disconnect the main pump motors during the energized condition, prior to transfer, and reconnect them after transfer when the residual voltage has been substantially reduced. A time delay of 2 to 10 minutes should be provided to prevent transfer back to normal source when normal service is recovered after an outage.

11.6.3 Water Level Sensors

The various types of pump control sensors, which regulate pump activity, include:

- Float switch
- Electrode probes
- Ultrasonic devices
- Tilting bulb with mercury switch
- Bubble-tube
- Entrapped air pressure switch

11.6.3.1 Float Switch

A float switch consists of a floating device attached to a moving tape or wire and pulley system. The float rises and falls with the water level in the sump. When the water level reaches a given height, the pumps are activated by a control switch and the pumps continue pumping water from the intake area until the water level drops. One particular specification of which the hydraulic designer should be aware is that float switches should not be used if the number of pump cycles is expected to exceed 4 per hour because they are typically relatively slow to respond. Also, the float will respond to fluctuations caused by turbulence in the sump. Thus designers will usually ensure that a minimum height of about 150 mm (6 inches) is used between successive start elevations of pumps and successive stop elevations of pumps.

11.6.3.2 Electrodes

Electrode probes are activated when the rise in water level causes an electric current to pass between the electrodes of the sensor thereby activating the pumps. The electrodes are suspended vertically from the top of the pit with their lower ends positioned at the level elevations at which the control device is to be actuated. These sensors can tend to corrode and can be fouled by floating debris and pollutants.

11.6.3.3 Ultrasonic Transducers

Ultrasonic devices emit a mechanical pulse that registers a change in the water level in the intake box. The sensor registers a change in frequency due to the rise in water level and the pump controls detect this change and activate the pumps.

11.6.3.4 Mercury Switch

A tilting bulb with mercury switch (Figure 11-6) registers a change in the orientation of the liquid mercury within the bulb caused by a change in the water level in the intake box. This type requires a significant change in level to open or close. Thus, the designer will usually apply a water level change of at least 150 mm (6 in.) between successive pump start elevations and between successive pump stop elevations.

Figure 11-6. Tilting bulb switch

11.6.3.5 Bubble Tube

A bubble tube regulates pump activity by registering a change in the orientation of the bubble within the bubble-tube caused by a change in the water level. A bubble-tube is placed inside a conduit installed between the wet well and dry well of dry-pit stations. The tube enters the wet well from the top and terminates in the dry well at an elevation above the highest possible water elevation. When the wet well and the dry well do not have a common wall then a bubble tube presents the best alternative.

11.6.3.6 Air Pressure Switch

The switch registers an increase in pressure due to the compression of the air in the inflow pipe resulting from an increase in the height of water in the wet well. When the water level reaches a given height and corresponding air pressure the switch is activated and the pump starts. When the water level drops, the switch closes and the pump stops.

11.6.4 Sensing and Alarm Systems

Sensors are provided for critical operating systems within the pumping station in order to detect changes in standard operating procedure. One such system is the gas detection system. Wet-pit stations with a large, closed, rectangular pump pit, require fuel presence alarms. Stations with a ventilated below-pavement storage unit, such as the dry-pit station, do not require a gas detection system.

Detection and alarm systems are typically used for:

- gas detection,
- heat detection,
- pump failure,
- pump usage monitoring,
- indicating high water, and
- electrical failure.

Information from the sensor (detector) can be made available at a remote point by telemetry.

11.7 MECHANICAL CONSIDERATIONS

11.7.1 Mechanical Packing

Packing is the term given to the materials used to control air and water leakage between the pump impeller casing and the dry side of the drive shaft casing. A natural or synthetic woven fabric is used. Since the shaft rotates in a fixed casing, the woven fabric must be lubricated. Typical packing materials include:

- flax,
- cotton,
- jute,
- graphite,
- rayon,
- carbon,
- metal strands or foils, and
- a variety of synthetics including Teflon^{TM}, TFE^{TM}, and Kevlar^{TM}.

Lubricants include:

- grease,
- graphite,
- oils,
- mica,
- silicone, and
- other synthetics.

The packing is usually selected by the pump manufacturer as part of the pump design or the project's mechanical designer. However, selection of the appropriate packing should be based on factors such as:

- pump operating pressure,
- shaft speed, and
- water temperature.

The designer will need to establish this information and relay it to whoever will select the packing.

11.7.2 Minimizing Motor Wear, Tear, and Overheating

A first consideration for minimizing potential problems is to keep the pump system as simple as possible. Submersible pumps are a popular choice because they are self contained and do not require long drive shafts and other couplings.

Motor wear, tear, and overheating can be minimized by reducing the friction between the moving parts within a mechanical system using adequate lubrication. Overheating may be a concern for:

anti-friction bearings. These may present a problem of overheating when too much lubricant (grease) is used, which causes the turning action of the rolling element to produce excessive fluid friction.

insulation windings. Damage to the insulation windings is the principal cause of engine failure. the starting motor, since the starting current can be as high as ten times the full-load operating current. Each start temporarily overheats the motor insulation, and the life of the motor is reduced by the number of starts. Normal motor overload protective relays will not protect a motor against damage from too many successive starts, so it is necessary to limit the number of starts over a given period of time. If the motor can come up to speed with little or no load, the heating effect is much less.

11.7.2.1 Bearings and Seals

Motor bearings, which are either plain or anti-friction, support and control the motion of a rotating shaft.

Plain bearings are either sleeve or thrust types that depend on a lubricating film to reduce friction between the shaft and the bearing. When properly designed and lubricated, plain bearings develop oil films that have tremendous load-carrying capabilities.

Anti-friction bearings operate on the principle of rolling contact between elastic circular bodies. The resistance of this rolling action is quite low. At low speeds, ball and roller bearings develop so little resistance through rolling that they are superior to plain bearings. Individual anti-friction bearings can support, at the same time, both radial and thrust loads in varying degrees -- a characteristic not typical of plain bearings.

The following information may be of assistance in the consideration and selection of bearings and seals for inclusion in the pump station design:

The life of ball or roller bearings depends on fatigue strength of the material and decreases as speed and load increases.

Rotating seals and other mechanical and chemical devices help protect against unfavorable environmental conditions, such as vibration, poor fits, corrosion or abrasive dirt.

Normally standard horizontal motors up through 93 kW (125 horsepower), 1800 rpm, are furnished with radial deep-groove ball bearings which permit the motor to be mounted in any position, including vertical, providing the downward thrust on the shaft is less than the weight of the motor.

Both ball and roller, as well as sliding element-type, are used to carry the weight of the motor rotating element and in some cases the weight of the pump shaft and impeller.

11.7.2.2 Lubrication

Lubrication provides the means whereby all the mechanical parts within a mechanical system may be kept moving and operating efficiently. Anti-friction bearings, which are relatively insensitive to viscosity changes, require only small amounts of lubrication. All vertical motors of 75 kW (100 horsepower) and over should have the thrust bearing and lower guide bearing oil lubricated, with visible means for checking oil level and quality. Motors below 75 kW (100 horsepower) may be grease lubricated, but the system must provide for flushing out and replacing old grease.

This page intentionally left blank.

12. CONSTRUCTION, OPERATION AND MAINTENANCE

12.1 INTRODUCTION

This chapter contains suggestions that should be considered by those responsible for the design, construction, operation and maintenance of highway stormwater pump stations.

12.2 CONSTRUCTION

12.2.1 Safety

Safety during construction includes:

- providing adequate workspace with buffer zones,
- providing traffic control during construction, which conforms to National or State manuals on uniform traffic control devices,
- maintaining stringent security to minimize the potential for vandalism and the injury of intruders, especially children.

12.2.2 Temporary Sediment Control

The designer should establish appropriate sediment and pollution control measures that the contractor must adhere to during construction. The primary criteria are the National Pollution and Discharge Elimination System (NPDES) permit requirements for industrial activities. The designer and contractor should make themselves familiar with the pertinent best management practices and notice of intent (NOI) procedures applicable in the particular state.

Even if disturbed areas are smaller than those established by NPDES criteria, currently 2 ha (5 acres), the designer should adopt the most practicable best management practices for the site conditions. Excessive sediment loads can overload the pump system, possibly causing premature failure. Therefore, in addition to typical perimeter controls, it is not excessive to consider temporary measures such as:

- sediment traps,
- sedimentation basins, and
- diversion berms.

Refer to the FHWA publication, *Urban Drainage Design Manual, Chapter 10*,[10] for a discussion on various measures. The control measures should be inspected frequently, especially after any rainfall that results in significant runoff.

12.2.3 Contract Issues

It is recommended that the highway agency does not accept ownership of the pump station until all of the following conditions have been met:

- construction of the entire pump station is complete,

- construction of all stages of construction that drain to the pump station are complete, and
- the pump station components have been field tested.

The pump station is likely to be used to drain the work site during staged construction. In addition to handling normal runoff, the pump station can be subjected to temporary excessive and destructive sediment and debris loads that may emanate from the construction site. Measures to limit sedimentation (discussed in Section 12.2.2 - Temporary Sediment Control) should be established and maintained throughout the construction. Testing is discussed in Section 12.3 - Pump Station Testing.

12.2.4 Architecture

Pump station architecture is usually only thought of in terms of visual appearance, which varies greatly from one area to another depending upon any prevailing architectural standards of the agency, station location and other factors. Interior arrangements of space should accommodate necessary pumping machinery, mechanical appurtenances and other pump station features including the minimal space and facilities required for safe access, safe working areas and the convenience of personnel.

12.2.5 Aesthetics

The aesthetic value of the pump house structure and the suitability of the design to its surroundings is becoming an increasingly important part of the pump station design process. The choice of materials for the pump station superstructure can enhance the aesthetic value of the structure.

The use of masonry may provide pleasing results. The masonry will usually be either brick or concrete block, although natural stone may be used, either as a solid wall or a veneer. A face brick of some type is usually aesthetically pleasing, with or without exposed concrete columns, beams or other features. Concrete block masonry is usually preferred over brick due to the greater variety of block type and size available.

12.2.6 Construction Materials

Reinforced concrete is the usual choice for the substructure of a pump station, but the method of placement varies. Cast-in-place construction is usually used for rectangular pump pits. Precast concrete units, which are normally fabricated either at the site or elsewhere and then sunk to the required level, are usually in the form of circular caissons. Cast-iron or pressed steel rings can be used for caisson-type stations. Precast or cast-in-place concrete is usually the most suitable material for stormwater pump station substructures.

12.2.7 Pump Well Requirements

When vertical pumps are used, each pump must be set in the wet well and its motor driver or its gear-head, if engine driven, must be above the pump room floor. For submersible and vertical pumps, an opening in the floor is required to permit removal of the pump and associated fixtures.

For submersible and horizontal pumps, the pump/motor must be securely attached to the concrete floor using a baseplate assembly. The baseplate should be securely embedded in the top of the concrete floor. Pump base plates are normally quite thick and the anchor bolts should be recessed to avoid the hazard of personnel tripping. Tripping hazards from electrical conduit stub-ups and runs or from plumbing and piping runs must be avoided throughout the pump room.

12.2.8 Access to Pump Equipment

The stormwater pump station house is an enclosing structure and must provide access to the pumps and other equipment. The enclosure must have doors, roof hatches, or covered openings through which equipment can be passed or debris can be removed with a mobile crane. Enough room should be incorporated into the planning and design of a pump station for the operation of the crane and hoist and removal of the pumps and heavy equipment.

12.2.9 Lifting Devices

Provision must be made in the design of a pump station for handling pumps and other equipment. For large pumps, a common approach is to use overhead, power-operated, bridge-cranes with lay-down or work-areas on the pump room floor where equipment can be dismantled or repaired. A more cost-effective alternative, however, is to provide simple hand-chain lifting devices or to use mobile cranes or hoists to remove equipment from the pump station for repair at a central location.

12.3 PUMP STATION TESTING

Pumps are subjected to numerous performance tests including factory tests and model tests in order to ensure correct pump operation.

Field testing should take place after the pumps have been installed at the station, prior to being accepted. Testing should:

- ensure correct operation of pumps,
- ensure operating efficiency of the station,
- check power supply,
- check pump performance, and
- check alarm system.

12.3.1 Preliminary Considerations

The accuracy and reliability of any field testing depends on:

- pump installation, which should be performed with the field testing specifically in mind so as to provide for the use of suitable calibrated instrumentation,
- reliable and precise instrumentation,
- proper installation of instrumentation, and
- responsible and qualified technical personnel to perform the testing.

12.3.2 Preliminary Testing

Instrument accuracy and reliability needs to be assured before proceeding with testing. The ability of the personnel assigned to perform the testing also should be assessed. A number of preliminary tests may be necessary to ensure that the recorded data is accurate and reliable.

Accurate and reliable data should fall within tolerances and fluctuations specified by the manufacturer and/or Hydraulic Institute Standards. Preliminary testing is performed to determine the following quantities:

- head differential across pump,
- discharge,
- suction,
- speed, and
- power input to pump.

Test tolerances for pumps, with respect to capacity, total head, or efficiency at the rated or specified conditions, require no negative tolerance or margin. Conformity with one of the following tolerances is required.

- At rated total dynamic head: +10% of rated capacity
- At rated capacity: +5% of rated total dynamic head under 150-m (450 ft.)
                          +3% of rated total dynamic head 150-m and over.

12.3.3 Testing

12.3.3.1 Pump Performance Tests

Pump performance tests should be made preferably at the maximum, average, and minimum total dynamic heads specified by the designer. The required test information for pump performance tests typically includes the following:

12.3.3.2 General

The pump tester should document the following information before each test is carried out:

- date,
- location,
- test number,
- name of person performing test, and
- pump identification number, if applicable.

12.3.3.3 Head

The following head determinations are made to check conformance:

- head at standard atmospheric pressure,
- gauge head, and
- total dynamic heads (maximum, average and minimum).

12.3.3.4 Capacity

The capacity test includes determining:

- volume of the storage unit, and
- rate of flow (volume/test period).

12.3.3.5 Power

The following quantities are required for each of the pumps in combination:

- volts (unloaded),
- volts (loaded),
- measured current,
- 100% full load current,
- adjustment factor, and
- adjustment current.

12.3.3.6 Performance Curves

Pump performance is a function of the following quantities:

- total dynamic head,
- brake power,
- net positive suction head, and
- derived pump efficiency.

The performance tests that determine these quantities are typically carried out over a range of flows at constant speed.

12.3.3.7 Inspection Tests

Visual inspection of the pumps is part of the testing procedure and is necessary prior to testing to ensure the proper orientation of the pump. The following are included usually in the inspection:

- alignment of pump and driver,
- direction of rotation,
- electrical connections,
- operation of stuffing boxes and lubrication systems, and
- wearing ring clearance.

12.3.3.8 Hydrostatic Tests

These tests are normally carried out by the pump manufacturer prior to pump installation. Each part of the pump should be capable of withstanding a hydrostatic test at test pressures that should be maintained for at least five minutes, at not less than the greater of the following:

- 150% of the pressure that occurs in the pump when operating at rated conditions, or
- 125% of the pressure that occurs in the pump when operating at rated conditions, but with the pump discharge valve closed.

12.4 OPERATION AND MAINTENANCE

12.4.1 General Considerations

The designer of a stormwater pump station should pay careful attention to the affect that design features have on operation and maintenance. The design should facilitate and minimize the cost of operation and maintenance without adding unreasonably to construction costs. The design also should minimize the possibility of malfunctions and stations becoming inoperable.

Primary design considerations or measures to help minimize operation and maintenance costs include:

- ensuring proper hydraulic design,
- providing for debris trapping and removal,
- providing for sediment handling or control and removal,
- considering appropriate pump type for given conditions,
- accounting for the experience of local maintenance personnel, and
- maintaining consistency with existing maintenance capabilities.

12.4.2 Safety

All elements of the pump station should be thoroughly reviewed for safety during operation and maintenance. Particular attention should be given to all safety features in order to maintain a safe working environment. All safety features included in the design and construction of pumping stations must be in accordance with the prescribed local, state and national safety laws and codes. Safety features include:

- ladders, stairwells, railings and other access points for use by maintenance and operations personnel in accordance with OSHA and local standards,
- guards and railings placed around all access holes and openings and all mechanical equipment with which the operator might come into contact including belt drives, gears, chain drives, rotating drive shafts and other moving parts,
- adequate space for the operation and maintenance of all equipment,
- warning signs near dangerous machinery and safety rules posted in appropriate locations throughout the station,
- paying particular attention to providing proper and reliable lighting especially when there is moving equipment,
- rubber mats provided in front of all electrical equipment where there is a possibility of electric shock,
- electrical equipment properly insulated and grounded,
- switches and controls of the non-sparking type,
- proper drainage to eliminate the possibility of slippery surfaces,
- a telephone to permit an operator to maintain regular contact with the main office,
- air testing and monitoring equipment in order to assure a safe environment within the station,
- ventilation systems to minimize the dangers of explosions and those arising from pollutants,
- fire extinguishers throughout the station in case of emergency,
- permanent traffic warning signs, and
- temporary traffic control, if access is limited.

An entry plan should be developed as part of the operation and maintenance procedures for each pump station. The plan should identify measures to be taken prior to and during any visit to the pump station, including monitoring of environmental conditions, especially air quality. All measures should conform with current OSHA requirements.

12.4.3 Access to the Pump Station Site

Personnel safety is a paramount concern, so access should not interfere with vehicular traffic. If right-of-way or other physical restrictions preclude construction of a dedicated service road or driveway, temporary barricades and signs should be placed during inspection and maintenance in conformance with the *National Manual for Uniform Traffic Control Devices*.[14]

12.4.4 Removal of Collected Debris and Sediment

Normal operation of the collection system will result in a build up of debris on trash racks, gratings and screens and sediment in the storage unit and sump. The build-up of debris may cause the bearings, impellers and bowls to be damaged reducing the pump efficiency and often resulting in failure.

An operation and maintenance schedule should identify a frequency of inspection to identify the need for debris and sediment removal.

12.4.5 Provision for Replacing Pumps

Pumping stations should be designed for easy maintenance and/or removal and replacement of pumps. The alternatives usually are a permanent hoist in the station or use of mobile hoisting equipment.

12.4.6 Typical Problems

The following is a list of design practices that can result in unnecessary maintenance requirements:

- poor siting of the station making access difficult,
- inadequate erosion control and slope stability measures producing excessive sediment and resulting in excessive wear of pump equipment,
- inadequate wet well dimensions causing vortexing or cavitation leading to excessive noise and wear,
- improper orientation of pumps causing vortexing, noise and wear, and
- poor trash handling resulting in clogging.

Generally, if the considerations and criteria discussed in this manual are followed, the potential for significant maintenance problems will be minimized.

12.4.7 Troubleshooting

Table 12-1 identifies some specific problems, their likely cause(s), and possible mitigation measures.

Table 12-1. Pump operation problems and typical solutions

Condition	Likely Cause	Mitigation/Avoidance
Frequent flooding of road low point	On and off sensor levels incorrectly set	• Adjust sensors to activate at design levels • Check sensors during pump testing
	Clogged intake	• Provide trash rack and frequent cleaning
	Inadequate pump capacity	• Add pump(s), • replace pumps, or • add storage unit
Rapid cycling	On and off sensor levels incorrectly set	• Adjust sensors to activate at design levels • Check sensors during pump testing
	Inadequate usable storage	• Increase storage unit capacity
	Excessive pump size	• Increase storage unit capacity, or • Add low flow pump
Excessive noise/vibration	Cavitation or vortexing	• Increase submergence and pump clearances, or • add baffles/vortex suppressors
	Incorrectly aligned pump	• Check installation and adjust alignment
Excessive wear on pumps	Rapid cycling, cavitation or vortexing	See rapid cycling and excessive noise/vibration
Overheating of pumps	Rapid Cycling	See rapid cycling
Sump pump failure	Low fluidity of sump sediment.	Provide flushing system

12.4.8 Inspection and Maintenance Schedule

The highway agency should develop a maintenance log that includes:

• the required frequency of inspection,
• an inspection check list, and
• routine maintenance measures.

12.4.9 Monitoring

By necessity, highway stormwater pumps must operate automatically. There are numerous components that may be subject to failure. With only one or two pump stations, it may be efficient enough to rely on local highway maintenance offices to manually monitor the condition of the stations. As the number of stations in an agency's inventory increase, it becomes feasible and desirable to use some kind of automated monitoring system. Typical features to monitor include:

- high water in the wet well,
- number of starts for each motor,
- cumulative operation time of each pump,
- leakage in dry well (for dry-pit stations),
- debris levels/amounts,
- sediment levels,
- motor/engine failure,
- ambient temperature,
- temperature of bearings and motor wiring,
- smoke,
- gases, and
- unauthorized entry.

The simplest type of automated warning system is a flashing light or siren that relies on locals to call in to a listed number. Preferred systems allow remote telemetering of any of the above alarms, or data, to a monitoring station.

This page intentionally left blank.

13. RETROFITTING EXISTING SYSTEMS

13.1 INTRODUCTION

Existing pump stations may become inadequate, undependable or troublesome, often requiring excessive maintenance. Some kind of retrofit may be necessary. The type of retrofit will depend on the cause of the deficiency or the extent of change. This chapter identifies various problems that relate to hydraulic deficiencies and methods to mitigate them.

13.2 IDENTIFICATION OF DEFICIENCY

Usually, retrofit of an existing station is needed because of proposed changes to the highway or as a result of continual or increasingly poor performance of a system. Maintenance records should provide some insight into the type and possible cause of the problem. In some cases, the cause may be reasonably easy to identify and relatively cheap to mitigate. However, oftentimes the cause is difficult to ascertain and solutions are likely to be expensive. Some circumstances may warrant developing small-scale models to hone in on the specific problem and test alternative solutions.

13.3 HIGHWAY MODIFICATIONS

If the highway is to be modified, the runoff rates and collection system may change. In some cases, the combination of age of the existing system and increase in load to the pump station may warrant a complete redesign of the system. However, the designer should evaluate the system to determine if non structural or less extensive structural measures than replacement are viable.

13.3.1 Increased Pumping Capacity

The most likely need associated with a significant modification to the highway is the need to handle a higher inflow rate and total volume. Consideration should be given to using submersible pumps. Typically, they require lower cycle times for the same capacity, thus the required usable storage is lower than many other pump types. This could help reduce the need to adjust the wet well structure or collection system size. Recent developments may make submersible pumps a viable alternative for consideration now. When the pump station was originally developed, adequate submersible pumps were generally not available.

Any need to increase the pump capacity will warrant an evaluation of the station performance using the design procedures detailed in Chapters 7, 8, and 9.

13.3.2 Increased Storage Capacity

It may be cost effective to provide additional available and usable storage and avoid the need to replace the pumping equipment and controls. The addition of a storage unit or supplemental collection line conduit need not require the removal or reinstallation of any pump equipment. However, it is likely that the sensor levels for the pump switching will require adjustment to optimize the use of the new storage. It will be necessary to perform mass curve routing and cycling checks, as discussed in Chapters 7 and 8, to check the station performance with the new storage.

13.3.3 Elimination of the Structure

The designer should consider the possibility of constructing a gravity drain. The economics of this alternative may have changed since the original construction.

13.4 POOR STATION LOCATION OR DISCHARGE CONDITIONS

The following table identifies some problems and potential solutions that may relate to poor station location or discharge conditions.

Table 13-1. Site problems and possible mitigation

Problem	Possible Mitigation
Silting due to poor erosion control	1. Stabilize exposed areas 2. Add sediment basin 3. Add storage unit with sediment trap and sludge pump 4. Add sediment flushing system
Silting due to slope failures	1. Reduce slopes 2. Protect slopes 3. Use retaining wall
Pump house flooded by high magnitude floods, but pumps operating properly	Move station to higher ground
Outfall backwater submerges discharge line	1. Improve outfall conditions to reduce backwater 2. Increase elevation of discharge line and check pump operation or replace pumps

13.5 HYDRAULIC-RELATED DEFICIENCIES OF PUMP STATIONS

The following table identifies typical hydraulic deficiencies of pump stations and possible causes.

Table 13-2. Possible causes of pump station deficiencies

Deficiency	Possible Cause(s)
Excessive pump cycling	Oversized pumps, insufficient storage, incorrect switching levels or sequence.
Excessive pumping noise	Cavitation or vortexing due to low submergence, poor flow alignment, high velocities in the wet well or low NPSH.
Excessive and frequent flooding	1. Clogging of pump intake 2. Poor outfall conditions 3. Insufficient pump capacity 4. Insufficient storage capacity 5. Worn impellers
Excessive pump wear	Cavitation, vortexing, poor maintenance, high sediment load
Excessive energy (electricity/fuel) requirements	Cavitation, vortexing, poor pump selection, leaks, damaged impeller
Frequent clogging of pumps	Excessive debris or sediment due to poor screening or poor flushing

Many of the above problems relate to the same cause. Other deficiencies may exist, such as non-compliance with current OSHA safety and NEMA standards, but are considered beyond the scope of this manual.

13.6 REPLACEMENT OF PUMPS

The following are typical reasons for replacing the pump units of a station:

- pump and driver have exceeded useful life,
- sump dimensions appear to be satisfactory, but pumps cannot handle desired capacity, and
- life cycle cost of new pumps is lower than operating cost of existing pumps.

In many cases it will be necessary to replace the pumps because of excessive wear. Any time that pump replacement is considered, the entire design procedures, detailed in Chapters 7, 8, and 9, should be addressed to ensure efficient design and determine what, if any, adjustments to the sump dimensions and switching may be necessary.

Figure 13-1 shows one of three existing electrical motor-drives for vertical shaft pumps in a confined station on US 59 in Houston. The pumps were more than 25 years old. The electrical control system was replaced and new submersible pumps installed. The lower cycling requirements of the submersible pumps compared with the existing pumps avoided the need to adjust the wet well structure.

Figure 13-1. Old motor drive for vertical pump to be replaced

13.7 LEAKAGE

Leakage can cause a significant reduction in efficiency and an increase in power requirements. Typical causes of leakage are worn bearings and seals, and faulty valves. Wear and tear might be attributed to the age of the system. However, there may be other causes such as:

• excessive vibration, friction and impact caused by cavitation or vortexing,
• hydraulic transients such as water hammer resulting from sudden pump operation or termination,
• inadequate or absent pressure release valves, and
• excessive sediment, debris loads, or corrosion.

Normal maintenance should ensure that the integrity of seals and bearings is maintained. Possible remedies for vortexing, cavitation, and excessive sediment or debris are discussed below. If water hammer is the likely cause, provision for slow-closing valves may reduce future problems by reducing the rate of pressure changes. The pump manufacturer should be consulted for recommendations on appropriate valve types and locations for the specific pumps used.

13.8 MODIFICATIONS TO PUMP SWITCHING

Rapid cycling will likely cause overheating and excessive wear of electric motors. Adjustments to the switching elevations and sequence may resolve the problem if the following are true:

- the pumps are not beyond repair or similar replacement pumps are intended,
- no serious flooding of the pump station has been experienced,
- there is significant total storage available, and
- the pumps are determined to be cycling to quickly.

Two specific mitigation measures are:

1. Increase the elevation at which the pumps switch on. This will require an evaluation of the station performance using the mass curve routing procedure discussed in Chapter 7.
2. Use an alternating switching scheme, which is the preferred design measure.

13.9 MODIFICATIONS TO WET WELL

Problems such as vortexing and cavitation may not require replacement of the pump unit unless there is excessive wear. Oftentimes, such problems are a result of incorrect sump dimensions and clearances. In order to determine if modifications to the sump dimensions and clearances will reduce the problems, it is necessary to compare recommended criteria with actual dimensions to identify any significantly deficient values. If any actual values are deficient, some adjustment will be appropriate. Depending on the extent of the deficiency, adjustments to the existing sump that avoid replacing or significantly modifying the wet well structure may suffice.

13.9.1 Vortexing

The following table provides some different mitigation measures for deficiencies in sump dimensions and clearances. The table also references appropriate figures. Generally, these measures may be combined.

Table 13-3. Mitigation measures for deficiencies related to vortexing

Condition	Mitigation Measure	Reference Figure
Inflow to pump distance too short	Disperse flow using baffles or vanes	Figure 13-2
Pumps aligned with inflow	Realign pumps perpendicular to inflow	Figure 13-3
Submerged vortices resulting from low submergence	Place cone under intake	Figure 13-4
Excessive vortexing due to low submergence depth	Add suction umbrella or bell extension and splitter	Figure 13-5
Backwall distance too large and submerged vortices present	1. Triangular backwall and base splitter 2. False backwall	Figure 13-6
Eddying in sump	Smooth all transitions and sharp protruding edges using grout or other materials	Figure 13-7
Sump velocities exceed 1 m/s (3 fps)	Place turning vanes on pump bowl	

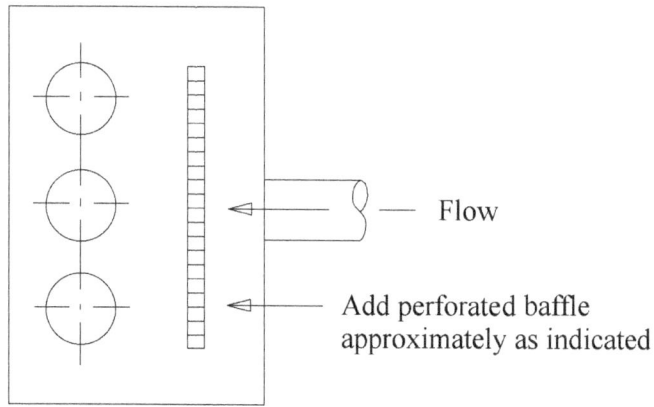

Figure 13-2. Addition of baffles in sump

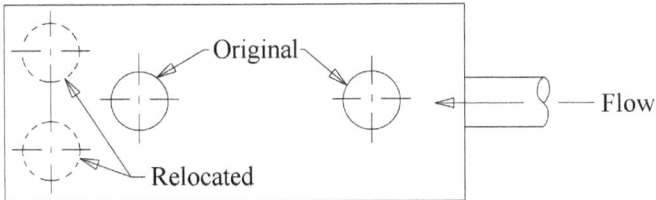

Relocate Pumps at Back Wall
as Indicated by Dashed Lines

Figure 13-3. Pump realignment

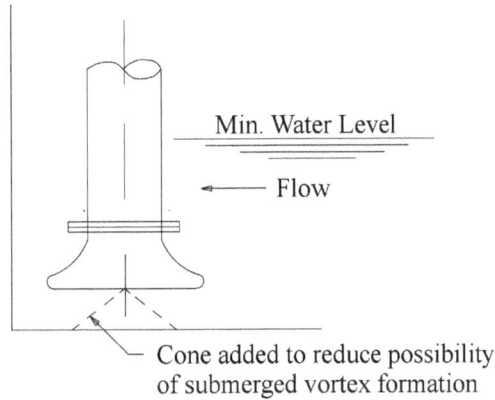

Figure 13-4. Addition of cone under intake

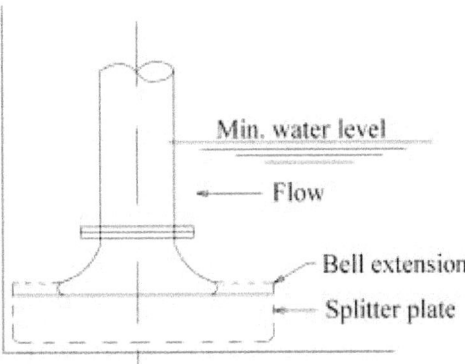

Dashed lines show enlarged bell and splitter
Splitter must be in-line with flow
Splitter is to prevent submerged vortexing

Figure 13-5. Addition of bell extension and splitter

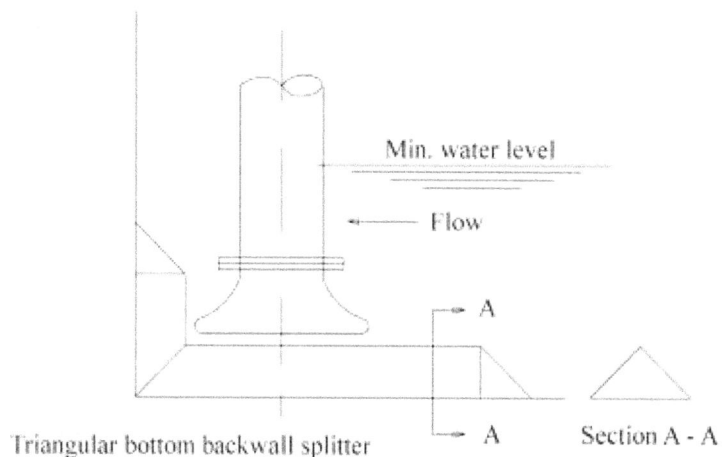

Figure 13-6. Addition of backwall splitter

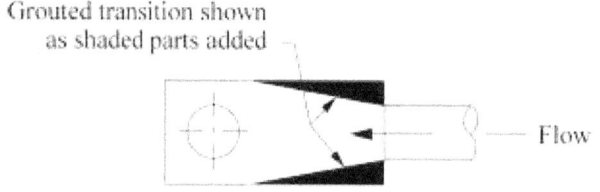

Figure 13-7. Addition of grouted transition

13.9.2   Cavitation

A primary cause of cavitation is insufficient net positive suction head available (NPSHA). The Hydraulic Institute suggests that minimum submergence values are based on minimizing the potential for vortexing. The net positive suction head required (NPSHR) is specific to a manufacturer's pump and the hydraulic conditions. If the NPSHA is lower than NPSHR, cavitation is likely. In fact, it is still likely under some conditions if NPSHA is greater than NPSHR. Therefore, a safety factor or margin is suggested. Non-structural mitigation options include increasing first pump-on elevation or changing pump to one with a lower NPSHR.

If the pump switching elevations are altered, it will be necessary to evaluate the station performance using the mass routing procedure described in Section 7.3.1 - Procedure to Perform Mass Curve Routing. This is also the case if the pumps are replaced with pumps of a different capacity to the original design.

13.9.3   Rapid Cycling

Rapid cycling may result from one or more of the following:

- small sump and collection system storage,
- first pump start elevation set too low, or
- pump capacity too large.

By inference, non-structural remedies may include increasing the first pump start elevation to achieve a usable storage that results in an adequate minimum cycle time, or adding a low flow pump.

13.9.4   Excessive Sediment, Debris and Corrosion

Excessive sediment or debris can clog the pump intake and erode or chip and fracture the impeller vanes. Additionally, corrosion can contribute to the degradation of the vanes, bearings and seals. Corrosion can be significant if the pump is continually immersed in sump water and is exacerbated by hazardous spills.

If minimal provision for screening of trash has been made, the obvious remedy is to improve the trash removal capability. The following are some specific suggestions:

- Use screened curb opening inlets or bicycle-safe grate inlets. Refer to the FHWA publication, *Urban Storm Drain Design*[10] for detailed discussion of inlet design.
- If existing trash rack allows clearance below first switch-on elevation, or between the top of the rack and the maximum water level in the wet well, replace the trash rack.
- If no specific schedule and guidance for inspection and maintenance exists, develop a plan and schedule and ensure that it is observed.
- If practicable, add automatic trash removal system.
- Minimize potential for trash buildup using street sweeping and trash collection.

If sediment build up is a problem, several possible remedial solutions exist. The preferred approach is to modify the sump to minimize horizontal floor area in the wet well, provide storage unit floor slopes of 2 % and allow the main pump to discharge sediment-laden water. Chapter 9, SUMP DIMENSIONS AND SYSTEM CHECKS provides discussion on design measures to accommodate sediment-laden water in rectangular and circular wells. The following are other alternatives:

- Add a self-flushing system and sludge/sump pump.
- Manually flush the sump and pump it dry with portable equipment after each significant storm or after a specified period, whichever is more frequent.
- Replace pumps with non-clog type pumps of similar performance characteristics.
- Provide a forebay to a lower sump where sediment can settle without interfering with the main pumps and add a sump pump.
- Provide a sediment trap upstream of the pump station.

Any of the above remediation measures that evacuate the sump will also help reduce the potential for corrosion of the impeller and bearings.

13.9.5 Replacing Wet Well

The remedies discussed in Vortexing (Section 13.9.1), Cavitation (Section 13.9.2), and Rapid Cycling (Section 13.9.3) are aimed at avoiding structural modifications. If actual sump dimensions are inconsistent with the recommended criteria, structural modifications may be necessary unless a pump replacement alternative is practicable. If substantial changes are made, it will be necessary to perform all the design procedures indicated in Chapters 6, 7, 8, and 9.

13.10 COMPARISON OF ALTERNATE METHODS

The cost of retrofitting a pump station may far exceed the original cost of construction. Therefore, the decision of what remediation measures to undertake should be based on:

- an economic evaluation of the viable alternatives, including a gravity drain,
- an appraisal of the added benefits that each alternative is likely to provide, and
- local design, construction, operation and maintenance experience.

14. REFERENCES

1. Stewart, Harry L., *Pumps*, revised by T. Philbin, Bobbs-Merill Co., Inc., 1984.

2. Manual for Highway Storm Water Pumping Stations Vol. 1, U. S. Department of Transportation, Federal Highway Administration, Report No. FHWA-IP-82-17-Vol.1, October 1982.

3. Dicmas, John L., *Vertical Turbine, Mixed Flow, and Propeller Pumps*, Mcgraw-Hill Inc. Book Company, 1987.

4. Volk, Michael W., *Pump Characteristics and Applications,* Volk & Associates, Inc. Oakland, California, Marcel Dekker Publishers, 1996.

5. McCuen, Richard H., Johnson, Peggy A. & Ragan, Robert M., *Highway Hydrology, Hydraulic Design Series No. 2,* Federal Highway Administration, Washington, D. C., U. S. Department of Transportation, Publication No. FHWA-SA-96-067, September 1996.

6. *Federal Aid Policy Guide*, Federal Highway Administration, Washington D.C.

7. *A Policy on Geometric Design of Highways and Streets,* American Association of State Highway and Transportation Officials (AASHTO), Suite 225, 444 North Capitol Street, N.W. Washington, D.C. 20001

8. *Technical Release 55, Urban Hydrology for Small Watersheds* (TR55), Natural Resource Conservation Service, 1981

9. *Technical Release 20, Hydrology* (TR 20), Natural Resource Conservation Service

10. Brown, S. A., Stein, S. M. & Warner, J. C., *Urban Drainage Design Manual, Hydraulic Engineering Circular No. 22,* U. S. Department of Transportation, Federal Highway Administration, Report No. FHWA-SA-96-078, November 1996.

11. Schall, James, D., & Richardson, Everitt, V., *Introduction to Highway Hydraulics, Hydraulic Design Series No. 4,* Federal Highway Administration, Washington, D. C., U. S. Department of Transportation, Publication No. FHWA HI 97-028, June 1997.

12. *Centrifugal Pump Design and Application,* Hydraulic Institute,

13. *Design of Rip Rap Revetment*, Hydraulic Engineering Circular, No. 11, 1989, Federal Highway Administration, Washington, D. C., U. S. Department of Transportation.

14. *National Manual for Uniform Traffic Control Devices*

15. American National Standard for Pump Intake Design, ANSI/HI 9.8-1998, Hydraulic Institute, New Jersey, December, 1998.

16. HY-22, Federal Highway Administration

17. Germain, James E., et al., *Design of Wastewater and Stormwater Pumping Stations*, Manual of Practice No. FD-4, prepared by Task Force on Pumping Stations under the direction of the Facilities Development Subcommittee Technical Practice Committee, Walter Pollution Control Federation, Washington D. C. 20037, 1981.

APPENDIX A: GLOSSARY

The following terms appear throughout this document and are defined as they relate to highway stormwater pump stations.

Term	Definition
Access hole	A hole through which one can access an underground element for repairs or inspections.
Affinity Laws	Rules that govern the performance of a centrifugal pump when speed or impeller diameter is changed.
Air Valve	A valve by which the entrance or escape of air can be regulated.
Available Storage	Total volume in storage system that is below the allowable highwater and above the lowest pumping level.
Backwall	The vertical surface behind a pump's intake bell.
Baffles	Obstructions to the flow that are aligned to distribute flow evenly between pumps.
Bay	A portion of the sump containing one pump.
Bearings	Motor bearings, which are either plain or anti-friction, support and control the motion of a rotating shaft.
Bell	The flared entrance to a pump or its suction pipe.
Best Efficiency Point	The discharge rate and total head at which the ratio of water power to brake power (hydraulic efficiency) is the highest it can be for the specific pump.
Brake Power	Brake power, BP (brake horsepower, BHP, in English units) is the actual amount of power required to be supplied to the pump to maintain the water power.
Caisson	A watertight, usually circular precast enclosure.
Cavitation	Hydraulic phenomenon in which vapor bubbles form and suddenly collapse (implode) as they move through a pump impeller. The formation of cavities between the back surface of an impeller blade and the liquid normally in contact with it.
Centrifugal Pump	A pump in which liquid is forced into the inlet side of the pump casing by atmospheric pressure or some upstream pressure. As the impeller rotates, liquid moves toward the discharge side of the pump, which creates a void or reduced pressure area at the impeller inlet. The pressure at the pump casing inlet, which is higher than this reduced pressure at the impeller inlet, forces additional liquid into the impeller to fill the void.
Check Valve	A pipe fitting used to prevent backflow to the pumps and subsequent recirculation.

Term	Definition
Close-coupled pump	Pump in which the motor and pump impeller share the same shaft. Also, the pump bearings and motor bearings are the same.
Closed impeller	The closed impeller has a shroud covering the vanes on the suction or front side and an axially oriented hub that provides the inlet for the liquid into the vane passageways.
Collection System	The system of conveyance elements that collect the stormwater and direct it to the pump station. The system usually includes channels, ditches, inlet lines and storm drain conduits.
Coupling	A fixture used to join two pipes. A typical coupling uses two flanged rings separated by a sleeve with rubber gaskets.
Start Elevation	Elevation at which a pump is set to begin operating.
Stop Elevation	Elevation at which a pump is set to cease operating.
Cycling	The starting, stopping then starting of a pump.
Design Capacity	The total stormwater runoff rate for which the pump station is designed. Also may refer to rate at which an individual pump will operate at the design point.
Design Point	The point of intersection of the system curve and the selected pump performance curve.
Diffuser (multi-volute)	A piece of casing, adjacent to the impeller exit, which has multiple passages of increasing area for converting velocity to pressure
Discharge	The rate of flow ejected from the pump
Discharge Line	A conduit through which the storm water exits the pump station.
Double suction	The action of a double suction impeller where the liquid enters the impeller from both sides.
Double volute	A hybrid between a single volute casing and a diffuser casing. Typically used in larger centrifugal pumps to reduce radial loads.
Dry well	The dry well of a dry-pit station contains the pumps and pumping accessories that discharge the storm water from the wet-well.
Dry-Pit Station	A pump station consisting of two chambers: a dry well and a wet-well. The water accumulates in the wet well and is pumped out by pumps installed in the dry well.
Elbow	A curved water passage by which water either leaves the pump (discharge elbow) or enters the pump (suction elbow).
Flap Gate	A circular device attached to the end of the discharge line to prevent backflow from the outfall to the discharge line.
Flap Valve	A check valve that uses a rubber flap
Flow line	The invert of a conduit
Force Main	A pressurized discharge line.

Term	Definition
Forebay	An area in the wet well between the storage unit and the individual pump bays.
Friction Loss	The energy loss in overcoming friction. Usually determined as an equivalent depth of water (head).
Gasket	Resilient material of proper shape and characteristics for use in joints between parts to prevent leakage.
Gate Valve	Devices used on collection lines to shut-off or permit water flow. The gate valve is not intended to regulate flow.
Grate Inlet	A lattice covering that permits flow to enter the storm drain conduit or storage unit.
Head	The measure of pressure expressed in equivalent height of water.
Hydraulic Grade Line	A line corresponding to the water level within a channel or a line that represents the pressure level in an enclosed conduit that is flowing full.
Hydraulic Institute	An entity consisting primarily of pump manufacturer's that establishes and publishes standards for the design and use of pumps.
Hydrograph	A representation of stormwater runoff versus time.
Impeller	The bladed member of the rotating assembly of the pump which imparts the principal force for the liquid pumped
Increaser	A transition from a smaller pipe diameter to larger pipe diameter
Inflow	A flowing in or into.
Inflow Mass Curve	A representation of cumulative inflow volume versus time.
Intake	The place at which a fluid is taken into a pipe, channel, etc.
Invert	The inside bottom elevation of a conduit or well at a specified location.
Manifold	A pipe terminal having several intake openings and a common discharge end.
Mass Curve Routing	The process of computing the volume of outflow as a function of the inflow volume, pumping rates, and storage.
Mass Inflow Curve	A curve representing the cumulative inflow volume with respect to time.
Mechanical Seal	A device used to help prevent leakage and reduce vibration, rotating friction and corrosion of the pump shaft.
Mixed Flow	A combination of radial and axial flow through an impeller
Net Positive Suction Head Required	Net positive suction head required (NPSHR) is the head above vapor pressure head required to ensure that cavitation does not occur at the impeller.

Term	Definition
Open impeller	The vanes of an open impeller are exposed on the suction or front side.
Operating Range	The range of total dynamic head over which the pumps in a system will operate.
OSHA	The Occupational Safety and Health Administration is an agency which establishes the safety requirements for pump stations.
Outflow	The storm water discharged by the pumps.
Packing	Material used to prevent leakage – usually a woven fiber and lubricant
Power Factor	The ratio of actual power to the apparent power consumed by an electric circuit.
Propeller	A propeller is a device consisting of a central hub with radiating blades used in low head, axial-flow, vertical pumps.
Pump	A device that increases the static pressure of a fluid. A pump adds energy to a body of fluid in order to move it from one point to another.
Pump Bowl	A vessel within which the impeller rotates and imparts energy to the water that is raised up the pump column.
Pump Column	The vessel above the pump impeller that conveys the discharge of a vertical pump to the discharge line.
Pump Cycling Time	Cycling refers to the time between starts of a given a pump. The shorter the cycling time, the more frequent a pump must start and stop.
Pump Discharge Curve	A representation of stage versus discharge for a system of pumps.
Pump House	An enclosed structure used to protect the pump control equipment, pump drivers and ancillary equipment.
Pump Performance Curve	A representation of the changing discharge rate with total dynamic head for a given pump.
Pump Sequencing	The order in which pumps are activated and deactivated
Pump Station	The collection of components used to lift highway stormwater runoff. A station includes the storage unit, wells, pumps, pump house and ancillary equipment.
Reducer	An element that is use to transition from one size pipe to a smaller one.
Retrofit	The replacement of various pump station components, typically the pumps.
Shaft	The cylindrical unit used to drive the pump
Single stage	A pump with only one impeller.

Term	Definition
Single volute	A spiral-like casing which directs the flow from the impeller at the center of the casing to the periphery. The shape of the volute converts a decrease in flow velocity at the periphery into a pressure increase.
Soffit	The inside top of a conduit.
Specific Speed	Specific speed is a dimensionless number expressed in terms of head per stage, pump capacity and shaft speed. Specific speed not only places an upper limit on the shaft speed for any particular combination of total head, flow, and suction conditions, but it also determines the form and proportions of the impeller.
Stage-Storage Curve	A curve that represents the storage that is available within the system at various stages.
Static Head	The change in elevation which the water must undergo. For typical stormwater pump stations, static head is normally measured between the water level in the sump and the higher of the water level at the end of the discharge side of the pump and the center of the end of the discharge pipe.
Storage	The provision for holding a volume of stormwater for the following purposes: ➢ reducing the size and number of required pumps, and ➢ ensuring minimum cycling times.
Storage Unit	A storage unit is a chamber that is provided to achieve a desired storage volume in excess of the storage provided by the collection system.
Stormwater	Rainfall runoff from the highway system and associated drainage area.
Stuffing Box	A packed casing through which the pump shaft extends that contains either a gland or mechanical seal to prevent leakage.
Submergence	Submergence is the static head of water required above the intake of the pump to prevent vortexing and cavitation.
Submersible Pump	A close-coupled pump and motor that are designed to be immersed. The motor is often encapsulated and filled with oil which is separated from the pumped liquid by a mechanical seal.
Suction Bell Velocity	Suction bell velocity is defined as flow divided by area measured at the outside diameter of the pump bell.
Suction Specific Speed	Suction specific speed is a non-dimensional index used to describe the geometry of the suction side of an impeller, or its NPSHR characteristics.
Sump	A basin in which the stormwater is collected and from which it is pumped out. Sometimes the terms wet well and sump are used interchangeably, though some wet wells may have a distinctly separate sump chamber.

Term	Definition
Sump pump	A sump pump also called an intake sump or sludge pump, is designed to remove the solids and sediment that are conveyed by the storm water through the inlet conduits into the storage box.
System curve	See System Head Curve
System Head Curve	A system head curve represents the variation in total dynamic head with pumping rate through the pumping system. At zero flow, the total dynamic head is equal to the total static head. As the pumping rate increases, the velocity head, friction losses, and pump losses increase. Thus, the total dynamic head increases with pumping rate.
Total Dynamic Head	Total dynamic head, TDH, represents the total energy required to raise the liquid from the intake to the discharge point.
Trash Rack	A trash rack is a screen that collects debris.
Type I Vortex	Type I vortexes are the start of vortex action in which tiny bubbles of air are pulled into the pump. These bubbles are not a significant impediment to pump performance.
Type II Vortex	Type II vortexes form for less than thirty seconds and pull air and floating debris into the pump. Pump capacity and horsepower are momentarily affected.
Type III Vortex	Type III vortexes are continuous and allow large amounts of air and debris into the pump accompanied by a sucking noise.
Umbrella	An attachment to the underside of the pump bell that is used to reduce the submergence requirements.
Ungula	The volume contained in a sloping conduit of circular cross section below a given horizontal surface
Usable Storage	The amount of storage that affects the pump cycling times. For any given pump, it is the volume contained between the pump's start and stop levels less any volume required to convey flow to the sump.
Valve	A device to regulate flow through a pipe
Vane	Any of several flat or curved pieces set around an axle and rotated about it moving air, water, etc.
Velocity Head	The head that represents the kinetic energy of flow, described by the term $V^2/2g$

Term	Definition
Volute	A spiral-shaped casing housing the impeller. The casing design assures an increase in the cross-sectional area of flow passages as the liquid moves through the casing from the impeller tip, thereby converting flow velocity to pressure.
Vortex	A circulation (swirling) of water from either the water surface or sump walls.
Water Power	Water Power WP (water horsepower WHP in English units) is the output power of a pump handling a given liquid at a given total dynamic head and discharge.
Wet well	A chamber of the pump station into which the storm water flows and from which it is pumped.
Wet-Pit Station	A pump station that contains a chamber (wet-well) into which the storm water is conducted. The water that accumulates in the wet-well is discharged by pumps that are installed in the wet well.
Wire-to-water power	Wire-to-water power is the total electrical power required to maintain the discharge and total dynamic head.

This page intentionally left blank.

APPENDIX B: ANNOTATED BIBLIOGRAPHY

American National Standard for Pump Intake Design, ANSI/HI 9.8-1998, Hydraulic Institute, Parsippany, New Jersey, October 1998.

A recent addition to the Hydraulic Institute standards, this standard provides guidance and recommendations for the design of a variety pump intakes for different applications. The standard adopts a different approach to recommended criteria for sizing intake structures by using pump bell or pump volute diameter as the primary basis for sizing the intake rather than discharge rate.

Anderson, Harold, *Centrifugal Pumps and Allied Machinery*, 4th Edition, Elsevier Advanced Technology, Elsevier Science Publishers Ltd., 1994.

This book discusses submersible pumps in general. Particular attention is paid to the advantages of submersible pumps when compared with horizontal pumps, motor and bedplate sets. Pump details, materials of construction, pump characteristics and thrust-bearing loadings are considered. The book also covers variable speed for centrifugal pumps and variable flow pumps.

Baumgardner, Robert H., *Hydraulic Design of Stormwater Pumping Stations: The Effect of Storage*, Transportation Research Record 948, National Research Council, National Academy of Sciences, Washington D. C., 1983, p83.

This paper provides design guidance for sizing storage facilities used to reduce peak rates of flow at storm water pumping stations. The effect of storage is evaluated by using a mass curve routing procedure that identifies the amount of storage required to reduce the peak rate of flow to the final discharge rate. The design procedure combines three independent components: the inflow hydrograph, the stage-storage relationship and the stage-discharge relationship.

Brennen, Christopher E., *Hydrodynamics of Pumps*, California Institute of Technology, Pasadena, California, Concepts ETI, Inc., and Oxford University Press, 1994.

This monograph discusses the fluid dynamics of liquid pumps. The focus is on the "special problems and design issues associated with the flow of liquid through a rotating machine." Particular attention is made to cavitation, which is the term used to describe the formation of vapor bubbles in regions of low pressure within the flow field of a liquid. Cavitation parameters and inception are discussed (Chapter 5), followed by chapters describing bubble dynamics, damage and noise (Chapter 6), pump performance (Chapter 7) and pump vibration due to cavitation (Chapter 8).

Brown, S. A., Stein, S. M. & Warner, J. C., *Urban Drainage Design Manual, Hydraulic Engineering Circular No. 22*, U. S. Department of Transportation, Federal Highway Administration, Report No. FHWA-SA-96-078, November 1996.

This report provides a comprehensive and practical guide for the design of storm drainage systems associated with transportation facilities. Pump stations are discussed in Chapter 9. The introduction to chapter 9 is followed by sections on pump station design considerations, design criteria, determining pump station storage requirements, design procedure and design philosophy.

Cheremisinoff, Nicholas P. and Paul N., *Pumps and Pumping Operations*, Prentice-Hall, inc., 1992.

This book is intended to assist the reader in making the proper choice, implementation and use of water pumps and their applications. The book is divided into six chapters. Chapter 1 considers flow dynamics and pumping principles. The chapter contains a brief overview of pumps, pumping operations, support equipment and classes of pumps. Chapter 2 covers pump classification and pumping services in greater detail. Chapter 3 discusses industrial pumps and their applications including installation and layout considerations. Oscillating displacement pumps are discussed in Chapter 4, and system analysis for pumping equipment is addressed in Chapter 5. The last chapter covers mechanical seals for pumps.

Dicmas, John L., *Vertical Turbine, Mixed Flow, and Propeller Pumps*, Mcgraw-Hill Inc. Book Company, 1987.

The intention of this book is to provide an overview of all aspects of vertical diffusion vane centrifugal, or turbine-type pumps in the range from 1,200 to 13,500 specific speed (0.44 to 5.00 type numbers), which are commonly referred to as vertical turbine, mixed flow and propeller pumps. The book provides a useful summary of standard pump specifications and construction features (Chapter 1) and definition of terms (Chapter 2). The mechanical design of vertical pumps is discussed in Chapter 3, followed by chapters on sump design (Chapter 6) and station design (Chapter 7). Further relevant topics for pump station design include operation procedures (Chapter 8) and vertical pump drives (Chapter 5).

Duan, Ning, *Optimal Reliability Based Design and Analysis of Pumping Systems for Water Distribution Systems*, Dissertation, University of Texas at Austin, August 1988.

This dissertation discusses the optimal reliability based design of pumps and storage tanks for water distribution systems. New methodologies are presented for the reliability analysis of pumping stations and the optimal reliability based design and analysis of pumping stations for water supply. Included is a reliability analysis of pumping station storage facility systems

(Chapter 4.3.2 and 4.4.2), tank design (Chapter 6.2-6.3) and reliability concepts such as failure and repair times (Chapter 2), all of which are important considerations in the design of pumping stations.

Garay, Paul N., *Pump Application Desk Book*, published by The Fairmont Press, Inc., 1990.

The objective of this book is to bring together the information necessary to select and apply pumps in systems of all kinds of fluids and purposes. The book limits itself to a consideration of only those factors that are necessary for an understanding of pump operation. Topics covered include pump types and arrangements (Chapters 2 - 4), pump performance, including NPSH (Chapter 11), pump operating principles and pumping system parameters. The book also considers the affects of design variation on use, economy and reliability.

Germain, James E., et al., *Design of Wastewater and Stormwater Pumping Stations*, Manual of Practice No. FD-4, prepared by Task Force on Pumping Stations under the direction of the Facilities Development Subcommittee Technical Practice Committee, Walter Pollution Control Federation, Washington D. C. 20037, 1981.

This important reference provides a practical overview of the wastewater and stormwater pumping station design process. It is intended as an outline of the process and as a remainder of the important considerations involved. The text is divided into nine chapters:

Chapter 1   Introduction
Chapter 2   Station Capacity
   Summary of the methods of estimating sanitary wastewater, stormwater and combined flows. Discussion of the hydraulic design of wet wells.

Chapter 3   Configuration and Design
   Discussion of location considerations, station types and structural and architectural design.

Chapter 4   Mechanical Design
   Compilation of the mechanical considerations of pump station design including installation, screening methods, pumping equipment, and heating and ventilation.

Chapter 5   Piping Systems
   Consideration is given primarily to the piping systems leading into and out of the pump station. Included is a discussion of the system and valve hydraulics.

Chapter 6   Electrical Design
   Contains a comprehensive description of the electrical equipment that must be selected and designed for pumping stations.

Chapter 7   Odor Control
        The origins and normal odor control practices are covered.

Chapter 8   Appurtenances
A discussion of some of the important accessory equipment in pumping station design including meters and gauges, seals, pump lubrication, hoists and ladders, aeration, drainage and safety features.

Chapter 9   Operation
Pump station maintenance and operation are considered.

***Hydraulic Institute Standards for Centrifugal, Rotary and Reciprocating Pumps*, 14th Edition, Hydraulic Institute, 1983.**

This standard reference is intended to "promote and further the interests of manufacturers of pumps, as well as the interests of the public in such matters as are involved in manufacturing, engineering, safety, transportation and other problems of the (pump) industry" (Article II, Objects, *By-Laws*). Included are chapters covering centrifugal, rotary and reciprocating pump standards, materials, and definitions for hardware terms and for slurry application.

Jung, Gunter and Schneider, Eugenia, *Design and Optimization of Seawater Pumping Stations for Industrial Plants*. A paper delivered at the International Conference on the Hydraulics of Pumping Stations, Manchester, England, 17 – 19 September, 1985. Organized and Sponsored by BHRA, The Fluid Engineering Centre, Paper 14.

This paper considers the site-specific aspects and number and types of pumps used at four pumping stations in the Middle East. Reliability and cost aspects are also discussed.

Karassik, Igor J., *Pump Handbook*, 2nd Edition, McGraw-Hill Book Company, 1986.

This handbook is intended to help in designing, selecting, operating and maintaining pumping equipment. The handbook discusses first the theory, construction details and performance characteristics of all the major types of pumps (Chapters 1 – 5). Pump drivers, couplings, controls, valves and the instruments used in pumping systems are discussed in Chapters 6 – 8. Pump services are covered in Chapter 9 and intakes and suction piping in Chapter 10. Chapter 12 discusses installation, operation and maintenance issues.

Lever, William F. & Associates, *Manual for Highway Storm Water Pumping Stations, Volumes 1&2*, Prepared for Federal Highway Administration, Washington D. C., U. S. Department of Transportation, Report No. FHWA-IP-82-17, October 1982.

This manual provides design information for highway storm water pumping stations. The manual is divided into the following chapters:

Chapter 1   Introduction
Chapter 2   Review of Current Practice
Chapter 3   Site Considerations
Chapter 4   Collection Systems
Chapter 5   Selection of Type of Station and Equipment
Chapter 6   Wet-Pit Design
Chapter 7   Dry-Pit Design
Chapter 8   Pumping and Discharge Systems
Chapter 9   Pumps for Stormwater Applications
Chapter 10   Electric Motors for Stormwater Pumps
Chapter 11   Engines and Accessories
Chapter 12   Electrical Systems and Controls
Chapter 13   Emergency Generators
Chapter 14   Construction Details
Chapter 15   Station Design Calculations and Layouts
Chapter 16   Retrofitting Existing Stations

Lobanoff, Val S. and Ross, Robert R., *Centrifugal Pumps: Design and Application*, 2nd Edition, published by Gulf Publishing Company in association with Plant Engineering, 1992.

The emphasis of this book is on hydraulic design and is divided into four parts. Below is a listing of selected topics discussed in each of the four parts:

Part 1   Elements of Pump Design

- Introduction
- Specific Speed and Modeling Laws
- Impeller Design
-           General Pump Design
- Volute Design
- Design of Multi-Stage Casing
- Double-Suction Pumps and Side-Suction Design
- NPSH

Part 2   Applications

- Vertical Pumps
- High Speed Pumps
- Double-Case Pumps
- Slurry Pumps
- Hydraulic Power Recovery Turbines
- Chemical Pumps

Part 3   Mechanical Design

- Shaft Design and Axial Thrust

- Mechanical Seals
- Vibration and Noise in Pumps

Part 4   Extending Pump life

- Alignment
- Bearings and Lubrication

Lohn, M. B., *The Hydraulics Design Aspects of Three Pumphouses*. A paper delivered at the International Conference on the Hydraulics of Pumping Stations, Manchester, England, 17 – 19 September, 1985. Organized and Sponsored by BHRA, The Fluid Engineering Centre, Paper 13.

This paper discusses the factors that determined the location of the pumphouses and the influence of this location on the pumphouse layout.

McCuen, Richard H., Johnson, Peggy A. & Ragan, Robert M., *Highway Hydrology, Hydraulic Design Series No. 2*, Federal Highway Administration, Washington, D. C., U. S. Department of Transportation, Publication No. FHWA-SA-96-067, September 1996.

This publication discusses the physical processes of the hydraulic cycle, applicable to highway drainage design engineers, and the methods that are used in the design of highway drainage structures. The hydrologic methods discussed in the manual include:

- frequency analysis for analyzing gauged data,
- empirical methods for peak discharge estimation,
- hydrograph analysis, and
- synthesis.

The peak discharge methods discussed include:

- regression equations,
- the rational method, and
- the SCS Graphical method.

Additional methods discussed in the manual include hydrologic methods used in arid lands, and detailed methods for the planning and design of detention basins

Murray, B. G., Ratcliffe, G. A. & Palmer, S., Development Engineering International Ltd., *Development in Pump Condition Monitoring*, a paper presented at the Eleventh Technical Conference of the British Pump Manufacturers' Association: Pump Technology: New Challenges – Where Next? Churchill College, Cambridge, England, 18 –

20 April, 1989. Published jointly by BHRA (Information Services), The Fluid Engineering Centre, Cranfield and Springer – Verlag, Paper 16.

The objective of this paper is to review some of the recent developments in machinery condition monitoring, with particular reference to pumping applications. Particular interest has been shown in three particular areas due to developments in instrumentation and computer equipment: Vibration Monitoring, Lubrication Oil Monitoring and Machine Performance Monitoring. The paper discusses recent developments in each of these three areas.

Pollak, F., *Pump Users Handbook (Third edition)*, extracts taken from an original text by H. Addison, published by Trade and Technical Press Ltd., England, 1988.

This handbook emphasizes the importance of selecting the correct pumping device based upon a correct interpretation of pumping requirements. Included are detailed accounts of the suction performance of pumps such as sump design (Sections 3 & 4). Mechanical seals (Section 5b), electric pump drives and motor starting (Section 8), pipeline systems and valves (Section 9) and health and safety concerns (Section 13) are also considered.

Sanks, Robert L., *Pumping Station Design*, Butterworth Publishers, 1989.

This is the only reference book available that deals specifically and comprehensively with the entire subject of the design of all phases of water and wastewater pumping stations. The first eleven chapters contain the prerequisites for effective design including hydraulics, piping, water hammer, electricity and a description of water pumps. Chapters 12 – 20 consider system design, including pump and driver selection and general piping layouts for water, wastewater, and sludge pumping. The last ten chapters discuss appurtenances such as instrumentation and design, heating and ventilation, noise and vibration, and a comparison of types of pumping stations and pumps. What follows is the pump station design process specified by chapter name as given in *Pumping Station Design*.

Chapter 1	Introduction
Chapter 2	Nomenclature
Chapter 3	Flow in Conduits
Chapter 4	Pipes and Fittings
Chapter 5	Valves
Chapter 6	Fundamentals of Hydraulic Transients
Chapter 7	Control of Hydraulic Transients
Chapter 8	Electrical Theory
Chapter 9	Electrical Design
Chapter 10	Theory of Centrifugal Pumps
Chapter 11	Types of Pumps
Chapter 12	Pump Selection
Chapter 13	Electric Motors

Schall, James, D., & Richardson, Everitt, V., *Introduction to Highway Hydraulics, Hydraulic Design Series No. 4,* **Federal Highway Administration, Washington, D. C., U. S. Department of Transportation, Publication No. FHWA HI 97-028, June 1997.**

This publication supercedes the previous HDS 4, *Design of Roadside Drainage Channels (1965)*. The manual provides an introduction to highway hydraulics and to hydrologic techniques applicable to small areas. The hydrologic techniques include the rational equation and an overview of regression equations. Also included are fundamental hydrologic concepts, conduit flow principles, and open channel and closed conduit applications along with examples. The manual provides an overview of highway drainage design and should be useful to designers lacking extensive drainage experience.

Schermel, A., *Multiple Pump Wet Well Intake and Pumping Station Design at Ontario Hydro-Nuclear Generating Stations.* **A paper delivered at the International Conference on the Hydraulics of Pumping Stations, Manchester, England, 17 – 19 September, 1985. Organized and Sponsored by BHRA, The Fluid Engineering Centre, Paper 16.**

The water intake, pumping station and discharge structure designs for three nuclear generating stations are described. Associated economical and environmental constraints are discussed.

Sinclair, M. R. & Bjorkander, G., *A New Approach to the Design of High Capacity Storm and Foul Pumping Stations,* **a paper presented at the Eleventh Technical Conference of the British Pump Manufacturers' Association: Pump Technology: New Challenges – Where Next? Churchill College, Cambridge, England, 18 – 20 April, 1989. Published jointly by BHRA (Information Services), The Fluid Engineering Centre, Cranfield and SPRINGER – VERLAG, Paper 5.**

This paper will show that the adoption of a larger number of lower capacity submersible pumps, installed directly in a wet wall, with individual siphonic discharge into a surge chamber, can result in significant reductions in capital plant, construction and energy costs, reduced hydraulic and electrical supply transients and improved flexibility of operation and maintenance procedures.

Stewart, Harry L., *Pumps*, revised by T. Philbin, Bobbs-Merill Co., Inc., 1984.

The purpose of this reference is to describe and provide a better understanding of the fundamental and operating principles of pumps, pump controls and hydraulics. Types of pumps considered are Centrifugal, Rotary, Reciprocal and Special Service Pumps. The chapters on hydraulics include discussions of hydraulic principles (Chapter 3), accumulators (Chapter 7), power tools (Chapter 9), cylinders (Chapter 10), control valve operations (Chapter 12) and fluids (Chapter 13).

United States Army, Corps of Engineers, *Texas – Cocodrie Pumping Plant, Red River Backwater Area, Tensas Basin, Louisiana Fish and Wildlife Mitigation Plan*. Communication from the Assistant Secretary of the Army (Civil Works), U. S. Government Printing Office, Washington D. C., 1979.

The volume is a compilation of various correspondences and reports submitted by the U. S. Army Corps of Engineers discussing the proposed construction of a water pump station in the Tensas Basin. The documents, which consider the environmental concerns raised by the construction of the pumping plant and provide a thorough review of the mitigation process and other related environmental issues, are a valuable source of information in considering a site for a pumping station.

Walker, Roger, *Pump Selection – A Consulting Engineer's Manual*, Ann Arbor Science Publishers, Inc., 1972.

This book is intended to facilitate the selection of pumping equipment for water supply systems. Section 1 is concerned with the selection of pumping equipment. Section 2 describes suction and discharge conditions including discussions of NPSH, TDH, cavitation and system head curves. Section 3 discusses specifications including shaft speed, performance curves, stuffing boxes, mechanical seals, valves and piping, and pump drivers. Section 4 considers maintenance requirements.

Warring, R. H., *Pumping Manual,* 7th Edition, Trade & Technical Press Limited, 1984.

The *Pumping Manual* is considered the best source for practical information and data on the selection, installation, operation, and maintenance of industrial pumps, in every field of application. The manual has three sections. Section 1 has a general introduction and discusses pump principles and pump performance. Section 2 is divided into five parts. The sections discuss pump types and characteristics, duty pumps, pump construction, pumping practice and pump ancillaries, respectively. Section 3 covers pump services and applications.